Putting **Essential Understanding** of
Functions
into **Practice**

in Grades 9–12

Robert N. Ronau
University of Cincinnati
Cincinnati, Ohio

Dan Meyer
Stanford University
Stanford, California

Terry Crites
Northern Arizona University
Flagstaff, Arizona

Terry Crites
Volume Editor
Northern Arizona University
Flagstaff, Arizona

Barbara J. Dougherty
Series Editor
University of Missouri
Columbia, Missouri

NATIONAL COUNCIL OF
TEACHERS OF MATHEMATICS

www.nctm.org/more4u
Access code: FNC14346

Copyright © 2014 by
The National Council of Teachers of Mathematics, Inc.
1906 Association Drive, Reston, VA 20191-1502
(703) 620-9840; (800) 235-7566; www.nctm.org
All rights reserved

Library of Congress Cataloging-in-Publication Data

Ronau, Robert N., 1948– author.
　Putting essential understanding of functions into practice in grades 9–12 / Robert Ronau, University of Louisville, Louisville, Kentucky, Dan Meyer, Stanford, California, Terry Crites, Northern Arizona University, Flagstaff, Arizona ; Terry Crites, volume editor, Northern Arizona University, Flagstaff, Arizona ; Barbara J. Dougherty, series editor, University of Missouri, Columbia, Missouri.
　　pages cm
Includes bibliographical references.
ISBN 978-0-87353-714-8
1. Functions—Study and teaching (Secondary) I. Meyer, Dan, author. II. Crites, Terry, author, editor. III. Dougherty, Barbara J., editor. IV. Title.
QA331.3.R663 2013
511.3'260712—dc23

2013034244

The National Council of Teachers of Mathematics is the public voice of mathematics education, providing vision, leadership, and professional development to support teachers in ensuring equitable mathematics learning of the highest quality for all students.

When forms, problems, or sample documents are included or are made available on NCTM's website, their use is authorized for educational purposes by educators and noncommercial or nonprofit entities that have purchased this book. Except for that use, permission to photocopy or use material electronically from *Putting Essential Understanding of Functions into Practice in Grades 9–12* must be obtained from www.copyright.com or by contacting Copyright Clearance Center, Inc. (CCC), 222 Rosewood Drive, Danvers, MA 01923, 978-750-8400. CCC is a not-for-profit organization that provides licenses and registration for a variety of users. Permission does not automatically extend to any items identified as reprinted by permission of other publishers or copyright holders. Such items must be excluded unless separate permissions are obtained. It is the responsibility of the user to identify such materials and obtain the permissions.

The publications of the National Council of Teachers of Mathematics present a variety of viewpoints. The views expressed or implied in this publication, unless otherwise noted, should not be interpreted as official positions of the Council.

Printed in the United States of America

Contents

Foreword vii

Preface ix

Introduction 1
    Pedagogical Content Knowledge 1
    Model of Teacher Knowledge 3
    Characteristics of Tasks 7
    Types of Questions 9
    Conclusion 10

Chapter 1
The Concept of a Function 11
    Introducing Functions Qualitatively: Car-Wash Scenario 13
    Common Misconceptions in Students' Thinking about Functions 17
    Developing a Robust Understanding of Domain and Range 22
    Conclusion 26

Chapter 2
Covariance 27
    Levels of Covariational Reasoning 27
    Supporting Local and Global Perspectives on Covariation 31
        Lake Depth 31
        Parking Fees 34
        Boat Rental 36
    Supporting Reasoning about Quadratic Rates of Change 39
        Falling Object 41
        Filling Bottle 43

Contents

    Sliding Ladder 45

    Interpreting the Covariance of Two Variables 49

    Conclusion 51

Chapter 3
Combining and Transforming Functions 53

    Combining Functions 55

        Addition of functions 56

        Subtraction of functions 58

        Multiplication of functions 61

        Division of functions 63

        A summative activity in combining functions 66

    Composing Functions 68

        Extending understanding: Car-wash coupons and discounts 69

        Further explorations in combining and composing functions:
        Prom Room Rental, Delivery Fees 74

    Transformations 75

        Issues of scale: Painting the mascot 76

        Transforming with coefficients or constants 80

        Applications for translations: Car-Wash Prices, Water's Height,
        Water's Surface Area 81

    Conclusion 88

Chapter 4
Graphs as Representations of Functions 91

    Addressing Misconceptions about Graphs 93

        Velocity of One Car 93

        Comparing Two Cars, Given Distance 96

        Comparing Two Cars, Given Speed 99

        Two Walkers 102

Assessing Understanding of Graphs 107

Conclusion 109

    Giving students a robust sense of function: Chapter 1 110

    Supporting emerging understanding of covariation: Chapter 2 110

    Making combining, composing, and transforming functions
        natural activities: Chapter 3 111

    Helping students interpret graphs of functions flexibly: Chapter 4 112

Chapter 5
Looking Back and Ahead with Functions 115

Building a Base for Functions in Kindergarten–Grade 8 115

Extending Understanding of Functions after Grades 9–12 123

Conclusion 137

Appendix 1
The Big Ideas and Essential Understandings for Functions 139

Appendix 2
Resources for Teachers 143

Appendix 3
Tasks 147

Function Wall 148

Function Finder 150

Lake Depth 151

Parking Fees 154

Boat Rental 157

Falling Object 161

Sliding Ladder 164

Water's Height in a Trough 167

Water's Surface Area on the End of a Trough 171

Contents

    Velocity of One Car 175

    Comparing Two Cars, Given Distance 178

    Comparing Two Cars, Given Speed.. 180

    Two Walkers.. 182

References 187

Accompanying Materials at More4U

    Appendix 1
    The Big Ideas and Essential Understandings for Functions

    Appendix 2
    Resources for Teachers

    Appendix 3
    Tasks

        Function Wall

        Function Finder

        Lake Depth

        Parking Fees

        Boat Rental

        Falling Object

        Sliding Ladder

        Water's Height in a Trough

        Water's Surface Area on the End of a Trough

        Velocity of One Car

        Comparing Two Cars, Given Distance

        Comparing Two Cars, Given Speed

        Two Walkers

Foreword

Teaching mathematics in prekindergarten–grade 12 requires knowledge of mathematical content and developmentally appropriate pedagogical knowledge to provide students with experiences that help them learn mathematics with understanding, while they reason about and make sense of the ideas that they encounter.

In 2010 the National Council of Teachers of Mathematics (NCTM) published the first book in the Essential Understanding Series, focusing on topics that are critical to the mathematical development of students but often difficult to teach. Written to deepen teachers' understanding of key mathematical ideas and to examine those ideas in multiple ways, the Essential Understanding Series was designed to fill in gaps and extend teachers' understanding by providing a detailed survey of the big ideas and the essential understandings related to particular topics in mathematics.

The Putting Essential Understanding into Practice Series builds on the Essential Understanding Series by extending the focus to classroom practice. These books center on the pedagogical knowledge that teachers must have to help students master the big ideas and essential understandings at developmentally appropriate levels.

To help students develop deeper understanding, teachers must have skills that go beyond knowledge of content. The authors demonstrate that for teachers—

- understanding student misconceptions is critical and helps in planning instruction;
- knowing the mathematical content is not enough—understanding student learning and knowing different ways of teaching a topic are indispensable;
- constructing a task is important because the way in which a task is constructed can aid in mediating or negotiating student misconceptions by providing opportunities to identify those misconceptions and determine how to address them.

Through detailed analysis of samples of student work, emphasis on the need to understand student thinking, suggestions for follow-up tasks with the potential to move students forward, and ideas for assessment, the Putting Essential Understanding into Practice Series demonstrates best practice for developing students' understanding of mathematics.

The ideas and understandings that the Putting Essential Understanding into Practice Series highlights for student mastery are also embodied in the Common Core State

Foreword

Standards for Mathematics, and connections with these new standards are noted throughout each book.

On behalf of the Board of Directors of NCTM, I offer sincere thanks to everyone who has helped to make this new series possible. Special thanks go to Barbara J. Dougherty for her leadership as series editor and to all the authors for their work on the Putting Essential Understanding into Practice Series. I join the project team in welcoming you to this special series and extending best wishes for your ongoing enjoyment—and for the continuing benefits for you and your students—as you explore Putting Essential Understanding into Practice!

Linda M. Gojak
President, 2012–2014
National Council of Teachers of Mathematics

Preface

The Putting Essential Understanding into Practice Series explores the teaching of mathematics topics in grades K–12 that are difficult to learn and to teach. Each volume in this series focuses on specific content from one volume in NCTM's Essential Understanding Series and links it to ways in which those ideas can be taught successfully in the classroom.

Thus, this series builds on the earlier series, which aimed to present the mathematics that teachers need to know and understand well to teach challenging topics successfully to their students. Each of the earlier books identified and examined the big ideas related to the topic, as well as the "essential understandings"—the associated smaller, and often more concrete, concepts that compose each big idea.

Taking the next step, the Putting Essential Understanding into Practice Series shifts the focus to the specialized pedagogical knowledge that teachers need to teach those big ideas and essential understandings effectively in their classrooms. The Introduction to each volume details the nature of the complex, substantive knowledge that is the focus of these books—*pedagogical content knowledge*. For the topics explored in these books, this knowledge is both student centered and focused on teaching mathematics through problem solving.

Each book then puts big ideas and essential understandings related to the topic under a high-powered teaching lens, showing in fine detail how they might be presented, developed, and assessed in the classroom. Specific tasks, classroom vignettes, and samples of student work serve to illustrate possible ways of introducing students to the ideas in ways that will enable students not only to make sense of them now but also to build on them in the future. Items for readers' reflection appear throughout and offer teachers additional opportunities for professional development.

The final chapter of each book looks at earlier and later instruction on the topic. A look back highlights effective teaching that lays the earlier foundations that students are expected to bring to the current grades, where they solidify and build on previous learning. A look ahead reveals how high-quality teaching can expand students' understanding when they move to more advanced levels.

Each volume in the Putting Essential Understanding into Practice Series also includes appendixes that list the big ideas and essential understandings related to the topic, detail resources for teachers, and present the tasks discussed in the book. These materials, which are available to readers both in the book and online at www.nctm.org/more4u, are intended to extend and enrich readers' experiences and possibilities for using the book. Readers can gain online access to these materials by going to the More4U website and entering the code that appears on the book's title page. They can then print out these materials for personal or classroom use.

Because the topics chosen for both the earlier Essential Understanding Series and this successor series represent areas of mathematics that are widely regarded as challenging to teach and to learn, we believe that these books fill a tangible need for teachers. We hope that as you move through the tasks and consider the associated classroom implementations, you will find a variety of ideas to support your teaching and your students' learning.

Acknowledgments

The authors thank their reviewers for their insightful responses to this volume. They are especially grateful to Jenny Simmons, National Board Certified Teacher, Saltillo High School, Saltillo, Missouri, for her thoughtful contribution.

Introduction

Shulman (1986, 1987) identified seven knowledge bases that influence teaching:

1. Content knowledge
2. General pedagogical knowledge
3. Curriculum knowledge
4. Knowledge of learners and their characteristics
5. Knowledge of educational contexts
6. Knowledge of educational ends, purposes, and values
7. Pedagogical content knowledge

The specialized content knowledge that you use to transform your understanding of mathematics content into ways of teaching is what Shulman identified as item 7 on this list—*pedagogical content knowledge* (Shulman 1986). This is the knowledge that is the focus of this book—and all the volumes in the Putting Essential Understanding into Practice Series.

Pedagogical Content Knowledge

In mathematics teaching, pedagogical content knowledge includes at least four indispensable components:

1. Knowledge of curriculum for mathematics
2. Knowledge of assessments for mathematics
3. Knowledge of instructional strategies for mathematics
4. Knowledge of student understanding of mathematics (Magnusson, Krajcik, and Borko 1999)

These four components are linked in significant ways to the content that you teach.

Even though it is important for you to consider how to structure lessons, deciding what group and class management techniques you will use, how you will allocate time, and what will be the general flow of the lesson, Shulman (1986) noted that it is even more important to consider *what* is taught and the *way* in which it is taught. Every day, you make at least five essential decisions as you determine—

1. which explanations to offer (or not);
2. which representations of the mathematics to use;
3. what types of questions to ask;
4. what depth to expect in responses from students to the questions posed; and
5. how to deal with students' misunderstandings when these become evident in their responses.

Your pedagogical content knowledge is the unique blending of your content expertise and your skill in pedagogy to create a knowledge base that allows you to make robust instructional decisions. Shulman (1986, p. 9) defined pedagogical content knowledge as "a second kind of content knowledge..., which goes beyond knowledge of the subject matter per se to the dimension of subject matter knowledge *for teaching*." He explained further:

> Pedagogical content knowledge also includes an understanding of what makes the learning of specific topics easy or difficult: the conceptions and preconceptions that students of different ages and backgrounds bring with them to the learning of those most frequently taught topics and lessons. (p. 9)

If you consider the five decision areas identified at the top of the page, you will note that each of these requires knowledge of the mathematical content and the associated pedagogy. For instance, when you introduce functions, the examples you use require that you consider your students' knowledge of graphing and equations and the way in which they interpret relationships between dependent and independent variables. Your knowledge of functions can help you craft tasks and questions that provide counterexamples and ways to guide your students in seeing connections across multiple algebraic and geometric topics. As you establish the content, complete with learning goals, you then need to consider how to move your students from their initial understandings to deeper ones, building rich connections along the way.

The instructional sequence that you design to meet student learning goals has to take into consideration the misconceptions and misunderstandings that you might expect to encounter (along with the strategies that you expect to use to negotiate them), your expectation of the level of difficulty of the topic for your students, the progression of experiences in which your students will engage, appropriate collections of representations for the content, and relationships between and among functions and other topics.

Model of Teacher Knowledge

Grossman (1990) extended Shulman's ideas to create a model of teacher knowledge with four domains (see fig. 0.1):

1. Subject-matter knowledge
2. General pedagogical knowledge
3. Pedagogical content knowledge
4. Knowledge of context

Subject-matter knowledge includes mathematical facts, concepts, rules, and relationships among concepts. Your understanding of the mathematics affects the way in which you teach the content—the ideas that you emphasize, the ones that you do not, particular algorithms that you use, and so on (Hill, Rowan, and Ball 2005).

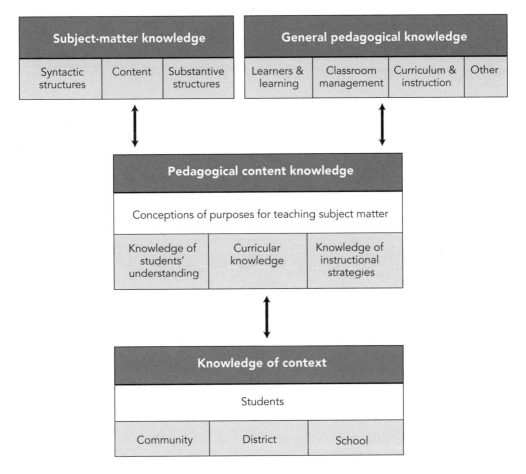

Fig. 0.1. Grossman's (1990, p. 5) model of teacher knowledge

Your pedagogical knowledge relates to the general knowledge, beliefs, and skills that you possess about instructional practices. These include specific instructional strategies that you use, the amount of wait time that you allow for students' responses to questions or tasks, classroom management techniques that you use for setting expectations and organizing students, and your grouping techniques, which might include having your students work individually or cooperatively or collaboratively, in groups or pairs. As Grossman's model indicates, your understanding and interpretation of the environment of your school, district, and community can also have an impact on the way in which you teach a topic.

Note that pedagogical content knowledge has four aspects, or components, in Grossman's (1990) model:

1. Conceptions of purposes for teaching
2. Knowledge of students' understanding
3. Knowledge of curriculum
4. Knowledge of instructional strategies

Each of these components has specific connections to the classroom. It is useful to consider each one in turn.

First, when you think about the goals that you want to establish for your instruction, you are focusing on your conceptions of the purposes for teaching. This is a broad category but an important one because the goals that you set will define learning outcomes for your students. These conceptions influence the other three components of pedagogical content knowledge. Hence, they appropriately occupy their overarching position in the model.

Second, your knowledge of your students' understanding of the mathematics content is central to good teaching. To know what your students understand, you must focus on both their conceptions and their misconceptions. As teachers, we all recognize that students develop naïve understandings that may or may not be immediately evident to us in their work or discourse. These can become deep-rooted misconceptions that are not simply errors that students make. Misconceptions may include incorrect generalizations that students have developed, such as the idea that the so-called vertical line test is a foolproof way of determining a function. These generalizations may even be predictable notions that students exhibit as part of a developmental trajectory, such as thinking that all straight lines are functions.

Part of your responsibility as a teacher is to present tasks or to ask questions that can bring misconceptions to the forefront. Once you become aware of misconceptions in students' thinking, you then have to determine the next instructional steps. The mathematical ideas presented in this volume focus on common misconceptions that students form in relation to a specific topic—functions in grades 9–12. This book shows how the type of task selected and the sequencing of carefully developed questions can bring the misconceptions to light, as well as how particular teachers took the next instructional steps to challenge their students' misconceptions.

Third, curricular knowledge for mathematics includes multiple areas. Your teaching may be guided by a set of standards such as the Common Core State Standards for Mathematics (CCSSM; National Governors Association Center for Best Practices and Council of Chief State School Officers 2010) or other provincial, state, or local standards. You may in fact use these standards as the learning outcomes for your students. Your textbook is another source that may influence your instruction. With any textbook also comes a particular philosophical view of mathematics, mathematics teaching, and student learning. Your awareness and understanding of the curricular perspectives related to the choice of standards and the selection of a textbook can help to determine how you actually enact your curriculum. Moreover, your district or school may have a pacing guide that influences your delivery of the curriculum. In this book, we can focus only on the alignment of the topics presented with broader curricular perspectives, such as CCSSM. However, your own understanding of and expertise with your other curricular resources, coupled with the parameters defined by the expected student outcomes from standards documents, can provide the specificity that you need for your classroom.

In addition to your day-to-day instructional decisions, you make daily decisions about which tasks from curricular materials you can use without adaptation, which tasks you will need to adapt, and which tasks you will need to create on your own. Once you select or develop meaningful, high-quality tasks and use them in your mathematics lesson, you have launched what Yinger (1988) called "a three-way conversation between teacher, student, and problem" (p. 86). This process is not simple—it is complex because how students respond to the problem or task is directly linked to your next instructional move. That means that you have to plan multiple instructional paths to choose among as students respond to those tasks.

Knowledge of the curriculum goes beyond the curricular materials that you use. You also consider the mathematical knowledge that students bring with them from grade 8 and what they should learn by the end of grade 12. The way in which you teach a foundational concept or skill has an impact on the way in which students will interact with and learn later related content. For example, the types of representations that you include in your introduction of functions are the ones that your students will use to evaluate other representations and ideas in later grades.

Fourth, knowledge of instructional strategies is essential to pedagogical content knowledge. Having a wide array of instructional strategies for teaching mathematics is central to effective teaching and learning. Instructional strategies, along with knowledge of the curriculum, may include the selection of mathematical tasks, together with the way in which those tasks will be enacted in the classroom. Instructional strategies may also include the way in which the mathematical content will be structured for students. You may have very specific ways of thinking about how you will structure your presentation of a mathematical idea—not only how you will sequence the introduction and development of the idea, but also how you will present that idea to your students. Which examples should you select, and which questions should you ask? What representations should you use? Your knowledge of instructional strategies, coupled with your knowledge of your curriculum, permits you to align the selected mathematical tasks closely with the way in which your students perform those tasks in your classroom.

The instructional approach in this volume combines a student-centered perspective with an approach to mathematics through problem solving. A student-centered approach is characterized by a shared focus on student and teacher conversations, including interactions among students. Students who learn through such an approach are active in the learning process and develop ways of evaluating their own work and one another's in concert with the teacher's evaluation.

Teaching through problem solving makes tasks or problems the core of mathematics teaching and learning. The introduction to a new topic consists of a task that students work through, drawing on their previous knowledge while connecting it with new ideas. After students have explored the introductory task (or tasks), their consideration of solution methods, the uniqueness or multiplicity of solutions, and extensions of the task create rich opportunities for discussion and the development of specific mathematical concepts and skills.

By combining the two approaches, teachers create a dynamic, interactive, and engaging classroom environment for their students. This type of environment promotes the ability of students to demonstrate CCSSM's Standards for Mathematical Practice while learning the mathematics at a deep level.

The chapters that follow will show that instructional sequences embed all the characteristics of knowledge of instructional strategies that Grossman (1990) identifies. One component that is not explicit in Grossman's model but is included in a model developed by Magnusson, Krajcik, and Borko (1999) is the knowledge of assessment. Your knowledge of assessment in mathematics plays an important role in guiding your instructional decision-making process.

There are different types of assessments, each of which can influence the evidence that you collect as well as your view of what students know (or don't know) and how they know what they do. Your interpretation of what students know is also related to your view of what constitutes "knowing" in mathematics. As you examine the tasks, classroom vignettes, and samples of student work in this volume, you will notice that teacher questioning permits formative assessment that supplies information that spans both conceptual and procedural aspects of understanding. *Formative assessment*, as this book uses the term, refers to an appraisal that occurs during an instructional segment, with the aim of adjusting instruction to meet the needs of students more effectively (Popham 2006). Formative assessment does not always require a paper-and-pencil product but may include questions that you ask or tasks that students complete during class.

The information that you gain from student responses can provide you with feedback that guides the instructional flow, while giving you a sense of how deeply (or superficially) your students understand a particular idea—or whether they hold a misconception that is blocking their progress. As you monitor your students' development of rich understanding, you can continually compare their responses with your expectations and then adapt your instructional plans to accommodate their current levels of development. Wiliam (2007, p. 1054) described this interaction between teacher expectations and student performance in the following way:

> It is therefore about assessment functioning as a bridge between teaching and learning, helping teachers collect evidence about student achievement in order to adjust instruction to better meet student learning needs, in real time.

Wiliam notes that for teachers to get the best information about student understandings, they have to know how to facilitate substantive class discussions, choose tasks that include opportunities for students to demonstrate their learning, and employ robust and effective questioning strategies. From these strategies, you must then interpret student responses and scaffold their learning to help them progress to more complex ideas.

Characteristics of Tasks

The type of task that is presented to students is very important. Tasks that focus only on procedural aspects may not help students learn a mathematical idea deeply.

Superficial learning may result in students forgetting easily, requiring reteaching and potentially affecting how they understand mathematical ideas that they encounter in the future. Thus, the tasks selected for inclusion in this volume emphasize deep learning of significant mathematical ideas. These rich, "high-quality" tasks have the power to create a foundation for more sophisticated ideas and support an understanding that goes beyond "how" to "why." Figure 0.2 identifies the characteristics of a high-quality task.

As you move through this volume, you will notice that it sequences tasks for each mathematical idea so that they provide a cohesive and connected approach to the identified concept. The tasks build on one another to ensure that each student's thinking becomes increasingly sophisticated, progressing from a novice's view of the content to a perspective that is closer to that of an expert. We hope that you will find the tasks useful in your own classes.

A high-quality task has the following characteristics:
Aligns with relevant mathematics content standard(s)
Encourages the use of multiple representations
Provides opportunities for students to develop and demonstrate the mathematical practices
Involves students in an inquiry-oriented or exploratory approach
Allows entry to the mathematics at a low level (all students can begin the task) but also has a high ceiling (some students can extend the activity to higher-level activities)
Connects previous knowledge to new learning
Allows for multiple solution approaches and strategies
Engages students in explaining the meaning of the result
Includes a relevant and interesting context

Fig. 0.2. Characteristics of a high-quality task

Types of Questions

The questions that you pose to your students in conjunction with a high-quality task may at times cause them to confront ideas that are at variance with or directly contradictory to their own beliefs. The state of mind that students then find themselves in is called *cognitive dissonance*, which is not a comfortable state for students—or, on occasion, for the teacher. The tasks in this book are structured in a way that forces students to deal with two conflicting ideas. However, it is through the process of negotiating the contradictions that students come to know the content much more deeply. How the teacher handles this negotiation determines student learning.

You can pose three types of questions to support your students' process of working with and sorting out conflicting ideas. These questions are characterized by their potential to encourage reversibility, flexibility, and generalization in students' thinking (Dougherty 2001). All three types of questions require more than a one-word or one-number answer. Reversibility questions are those that have the capacity to change the direction of students' thinking. They often give students the solution and require them to create the corresponding problem. A flexibility question can be one of two types: it can ask students to solve a problem in more than one way, or it can ask them to compare and contrast two or more problems or determine the relationship between or among concepts and skills. Generalization questions also come in two types: they ask students to look at multiple examples or cases and find a pattern or make observations, or they ask them to create a specific example of a rule, conjecture, or pattern. Figure 0.3 provides examples of reversibility, flexibility, and generalization questions related to functions.

Type of question	Example
Reversibility question	Corey wrote an equation whose graph contains the point (−3, 4). Write an equation that could be the one that Corey wrote.
Flexibility question	Toni graphed $2x = y - 3$. Tom graphed the equation $y = 4x - 3$. How does Tom's graph compare with Toni's graph?
Flexibility question	Teri said that $y > 3x - 1$ would have solutions in all quadrants except quadrant III. Do you agree or disagree with Teri? Explain your answer.
Generalization question	Draw a line with a slope of −2. Draw 3 more lines with a slope of −2. What do you notice about the lines?
Generalization question	Write an equation of a line that will pass through quadrants I, III, and IV.

Fig. 0.3. Examples of reversibility, flexibility, and generalization questions

Conclusion

The Introduction has provided a brief overview of the nature of—and necessity for—pedagogical content knowledge. This knowledge, which you use in your classroom every day, is the indispensable medium through which you transmit your understanding of the big ideas of the mathematics to your students. It determines your selection of appropriate, high-quality tasks and enables you to ask the types of questions that will not only move your students forward in their understanding but also allow you to determine the depth of that understanding.

The chapters that follow describe important ideas related to learners, curricular goals, instructional strategies, and assessment that can assist you in transforming your students' knowledge into formal mathematical ideas related to functions. These chapters provide specific examples of mathematical tasks and student thinking for you to analyze to develop your pedagogical content knowledge for teaching functions in grades 9–12 or to give you ideas to help colleagues develop this knowledge. You will also see how to bring together and interweave your knowledge of learners, curriculum, instructional strategies, and assessment to support your students in grasping the big ideas and essential understandings and using them to build more sophisticated knowledge.

Students entering high school have already had some experiences that affect their initial understanding of functions. In addition, they have developed some ideas about functions at earlier grade levels. Students in middle-grades classrooms frequently demonstrate understanding of mathematical ideas related to functions in a particular context or in connection with a specific graph, equation, or table. Yet, in other situations, these same students do not demonstrate that understanding. As their teacher, you must understand the ideas that they have developed about functions in their prior experiences so you can extend this knowledge and see whether or how it differs from the formal mathematical knowledge that they need to be successful in reasoning with or applying functions. You have the important responsibility of assessing their current knowledge related to the big ideas of functions as well as their understanding of various representations of functions and their power and limitations. Your understanding will facilitate and reinforce your instructional decisions. Teaching the big mathematical ideas and helping students develop essential understandings related to functions is obviously a very challenging and complex task.

Chapter 1
The Concept of a Function

Essential Understanding 1*a*
Functions are single-valued mappings from one set—the *domain* of the function—to another—its *range*.

Essential Understanding 1*b*
Functions apply to a wide range of situations.

Essential Understanding 1*c*
The domain and range of functions do not have to be numbers.

The first big idea about functions presented in *Developing Essential Understanding of Functions for Teaching Mathematics in Grades 9–12* (Cooney, Beckmann, and Lloyd 2010) is that a function is a single-valued mapping from one set—the *domain* of the function—to another—its *range* (Carlson et al. 2002; Monk 1992; Vinner and Dreyfus 1989). Functions are a prominent feature of secondary mathematics, as reflected in the Common Core State Standards for Mathematics (National Governors Association Center for Best Practices and Council of Chief State School Officers 2010), where the function strand begins in grade 8 and continues throughout high school. But functions have multidisciplinary utility apart from any content standards. Pause in your reading and turn to the next page to respond to the questions in Reflect 1.1, which ask you to consider the usefulness of the concept of a function.

Functions are essential in every field of applied mathematics. They are useful to the statistician who deploys a set of probability functions that tells whether a certain outlying observation is significant or expected. The climatologist has a single-valued mapping of a given year to the global mean temperature of that year. Government accountants have single-valued mappings of a given year

> ### Reflect 1.1
>
> **In what mathematical fields, careers, or real-life events is the concept of a function useful?**
>
> **Under what circumstances is it useful to define a relationship between two sets as a relationship?**

to the amount of revenue that the government should expect to take in under current tax policy. In both of these last cases, if the climatologist's or the accountant's model for temperature or revenue returned *more than one* value for a given year, it would not be a function—and it would not be very useful.

Computer programmers rely on functional language to such an extent that an ascendant programming paradigm is labeled "Functional Programming." A graphics programmer might begin a function with the following line of code:

```
def drawASquare(topLeftXCoordinate,topLeftYCoordinate) {}
```

This example shows the utility of *multivariable* functions, functions where the domain includes *more than one* variable—in this case, both the x- and the y-coordinate of the top left corner of the square. After some difficulty, the programmer might realize that this function does not actually map to a single square for each pair of domain values. She would realize that the square must also be defined by some measure of *size*—a side length, for instance—not just position, and she could re-form the function with this line of code:

```
def drawASquare(topLeftXCoordinate, topLeftYCoordinate, squareSideLength) {}
```

In addition to the enormous value of functions in a variety of applied fields, they can simply be intellectually stimulating. When students see a set of domain values mapping to a set of range values, they may be inspired to wonder what kinds of functions could possibly account for that transformation. In the teaching of applied and pure mathematics, framing discussions related to domain in terms of destruction can engage students as they consider, for example, "What numbers would *break* this function?"

Introducing Functions Qualitatively: Car-Wash Scenario

Classroom discussions about functions and their applications will not be fruitful unless teachers can anticipate students' misconceptions, account for them, and take steps to eliminate them. This chapter will illustrate prevalent misconceptions related to the concept of a function and ways to help students overcome them.

To begin, meet Derek and Marta, two high school students who are organizing a car wash for a service club at their school.

> Derek and Marta are co-presidents of the Community Action Club at Grover Cleveland High School. Their club wraps up its school-year activities with a daylong car wash that raises funds for the next year's activities in the community. The car wash is growing in popularity and scope, with club members providing musical entertainment for drivers while their cars are cleaned.
>
> Derek and Marta are organizing this year's car wash. Their work involves locating a site, scheduling volunteers, advertising, and buying supplies—the soap and sponges without which the car wash would be nothing more than a line of dirty cars on a hot spring day. But club members encounter the same problem every year: They do not budget the right amount of money for supplies. Last year they ran out of soap in the middle of the car wash. They had to shut the car wash down and turn away many customers while Derek ran to the store to buy any soap he could find. The year before last, they bought *too much* soap, which is not as bad as buying too little, but still irritating. *They want a better method for predicting the amount of money they should spend on supplies every single year.*
>
> At a club meeting, they bring the question to their club adviser, Mr. Ramirez, who is a mathematics teacher at the school. Derek and Marta hope that he will just take the matter off their hands and make it his responsibility. Instead, he says, "You guys should use a function. I bet that would help."
>
> Derek and Marta trade confused looks, and Marta asks, "What's a function?"

Before reading further, consider what you would do at this moment in Mr. Ramirez's place. Use the questions in Reflect 1.2 to guide your thinking.

Putting Essential Understanding of Functions into Practice in Grades 9–12

Reflect 1.2

How would you explain the concept of a function to Marta and Derek without resorting immediately to heavy abstractions or unfamiliar vocabulary?

What metaphor might be useful to you here?

In what follows, note that Mr. Ramirez answers Marta's question carefully, closely monitoring the responses from Marta and Derek to each piece of information before proceeding:

"A function is a mathematical tool to help us manage how two values change together," Mr. Ramirez replies to Marta. "One of your values is changing—the amount you need to spend on supplies. We need to figure out another value that is changing with it and then find a function to describe how they change."

Neither Derek nor Marta has a clear idea yet of the definition of a function, so Mr. Ramirez explains: "Let's look at *relationships* between two values. A function is just a special kind of relationship." He pins four letters at intervals across the back wall: A, B, C, and D. He turns to the members of the club. "Okay, if your ride to school takes ten minutes or less, stand under A." Derek, Jamie, and Angela go to stand under A. "If your ride to school takes *longer* than ten minutes, stand under D." Denitha and Marta stand under D.

"Look at this," Mr. Ramirez says as he starts drawing on the board. "Our two sets of values are *your names* and *where you stand*." He writes all the students' names and then draws arrows to the letters under which they are standing (see fig. 1.1).

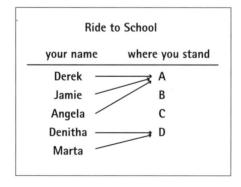

Fig. 1.1. An example of a function

Mr. Ramirez continues, "One thing you should notice here is that *none of you was confused*. Each of you had one place to go, and exactly one place to go. Let's look at another relationship. If you are wearing any shade of blue, stand under A."

Everyone but Denitha is wearing some kind of blue jeans, so all the club members start moving toward A. Mr. Ramirez then says, "If you are wearing white, stand under B. If you are wearing red, stand under C. If you are wearing black, stand under D."

Everyone freezes mid-stride. Marta says, "Mr. Ramirez, I'm wearing blue, white, and black. Where am I supposed to go?"

Mr. Ramirez says, "That is exactly my point. We have a *relationship*, but it is not a *function*." He starts drawing. "See, Marta, you're in the *domain* of the relationship: the input. But you don't know where to go in the *range* of the relationship: the output. You're confused. You have too many options. It's still a relationship, but it isn't a function. In a function, values in the domain go to exactly one value in the range." (Fig. 1.2 shows the situation created by Mr. Ramirez, with an input with more than one output.)

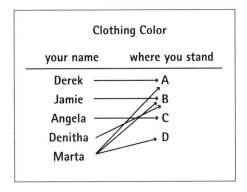

Fig. 1.2. An *input* with more than one *output*

"The rest of you have similar issues in this relationship, but even if you were each wearing exactly one color, it wouldn't matter. Marta's situation alone makes this a relationship, not a function." (Fig. 1.3 shows the relationship that Mr. Ramirez has set up—a relationship that is a not a function.)

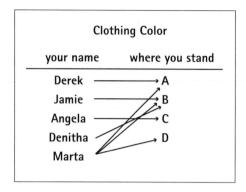

Fig. 1.3. A non-example of a function

"Let's try one more," Mr. Ramirez says. "You tell me if it's just a relationship or if it's a function. OK, here we go: If your birthday is in January, stand under A. If your birthday is in February, stand under B. If your birthday is in March, stand under C. If your birthday is in April, stand under D." (Fig. 1.4 maps the students' actions in response to Mr. Ramirez's instructions.)

```
                    Birth Month
   your name           where you stand
   Derek                    A
   Jamie                    B
   Angela  ─────────────→   C
   Denitha                  D
   Marta
```

Fig. 1.4. Every *input* needs an *output*

Only Angela, whose birthday is March 15, goes to stand at the wall. Everyone else remains seated. "Function or not?" asks Mr. Ramirez.

"Well, no one is going to more than one location, like last time," says Marta. "But some of us have *no* place to go. That means it's not a function. In a function, everything in the domain goes to exactly one place in the range."

"Bingo," says Mr. Ramirez. "If you can understand functions, you will be able to figure out how many supplies to buy for your car wash."

After pausing to let the students think about this, Mr. Ramirez poses a question: "So with our car wash, what are the different relationships we can examine, and which of them would be functions? Which function would be most relevant to our current challenge of buying supplies?"

Mr. Ramirez's Function Wall activity illustrates the kinds of conversations and confusions that can arise when students try to comprehend the concept of a function. Notice that Mr. Ramirez first introduces the notion of a function in response to a concrete need that the students in his service club have. Derek and Marta have a problem—the over- and under-budgeting of money for supplies for the car wash— and Mr. Ramirez suggests using a function to aid in determining a solution. Not every student will have that kind of practical problem, however, so Mr. Ramirez positions his other students to experience a perturbation—a moment of cognitive conflict. The instructions that he gives his students for standing along the wall are sometimes easy to carry out (under the letter for the time for riding to school) and sometimes very difficult (under the letter for the color of clothes).

At that point, the students lack any kind of formal vocabulary to explain to themselves the difference between the easy and difficult situations, so Mr. Ramirez takes the opportunity to define important terms. In a difficult situation, he explains, "You are confused. You have too many options. It's still a relationship, but it isn't a function. In a function, values in the domain go to exactly one value in the range." Use the questions in Reflect 1.3 to consider your own students' thinking.

Reflect 1.3

What misconceptions do your students commonly have about functions?

What examples and counterexamples can you present to help them confront those misconceptions?

Common Misconceptions in Students' Thinking about Functions

In a study of 152 preservice teachers and their conceptions of functions, Ruhama Even (1993) hints at the importance of the type of qualitative introduction that Mr. Ramirez gives his students to functions. In her interviews, Even detected three trends in students' thinking, the first of which was to consider that "functions are (or can always be represented as) equations or formulas" (p. 104). One study participant said, "A function is really an equation" (p. 105). Another said, "I think you could write all functions in terms of equations. It might be a trigonometric equation, like sin x, but in every term the y-value is going to be equal to some operation with x-value" (p. 105).

But functions can have many representations, of which equations are just one. And some functions, like the mapping of each student to a place identified by a letter along the wall, lack an equation or an algebraic representation altogether. Over the course of their secondary education, students see algebraic representations of functions more often than representations of any other type—so much so in fact that they risk conflating equations and functions completely. For that reason, it is productive to introduce functions non-algebraically, at first highlighting the single-valued mapping from domain to range in whatever form the function takes.

Three more misconceptions may arise as students develop their understanding of concepts related to functions:

1. They may think that the range must map back onto the domain with single values.

2. They may misunderstand how restrictions on the domain of a relationship affect the relationship.

3. They may rely on superficial or misleading indicators, such as the so-called vertical line test, to determine whether they are in fact dealing with a function.

It is useful to consider each of these sources of misunderstanding in turn.

Students may think that the range must map back onto the domain with single values. When the range of a function does in fact map back onto the domain with single values, the function is a special case called a "one-to-one" function. Just as a function is a special kind of relationship and therefore has characteristics and capabilities that distinguish it from a relationship (for one thing, a function allows single-valued predictions, whereas a relationship might return many values), so too does a one-to-one function have capabilities that another function that is not one-to one does not. (The inverse of a one-to-one function, for instance, is also a function, whereas the inverse of another function might not result in another single-valued function.) Later sections highlight this distinction, but for now, it is enough to say that it is important for students to understand which qualities of a relationship sufficiently describe a function and which qualities describe something more specific than a function.

The Function Wall activity is an attempt to confront this misconception. The full activity, which appears in Appendix 3, includes a task not used by Mr. Ramirez in the vignette: students who are shorter than five feet tall are to stand under A, students who are between five and six feet tall are to stand under B, students who are between six and seven feet tall are to stand under C, and students who are taller than seven feet are to stand under D. If you use this task with your students, they may laugh at this last request. They may also think that this relationship between students and their height is not a function, because no one is standing under D. (If someone *is* standing under D, you should of course create a section E for students who are taller than eight feet. You should also contact Rick Pitino, c/o the basketball program at the University of Louisville, at your earliest convenience.) Ask your students, "Do you feel confused about where you should stand? Do you feel any internal confusion at all?" If the students' response is no, say, "Then connect that clarity with the definition of a function. If all of you in the domain have exactly one place to stand, then you should call the relationship between all of you in the domain and the places to stand in the range a *function*."

Students may not understand how restrictions on the domain of a relationship affect a relationship, sometimes in effect turning a multi-valued mapping into one that is single-valued. Restricting the domain of a relationship to produce a function is an exercise of creativity in mathematics and an opportunity not to be missed. Restricting the domain of the relationship between students and the color of the clothes they are wearing to the two students in the class who are wearing *only one color* turns a relationship into a function. The capacity to do this may seem trivial at first, but it can be very important. Consider the graph of $y = \sqrt{x}$, shown in figure 1.5.

Putting Essential Understanding of Functions into Practice in Grades 9–12

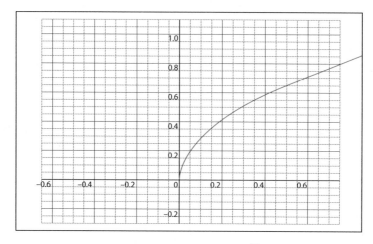

Fig. 1.5. The graph of $y = \sqrt{x}$

Does this graph show a function? A student who has a particularly rigid understanding of a function may say, "This isn't a function. Negative numbers don't have anywhere to go to in the range." In this case, you need to ensure that the student sees that *restricting* the domain creates a function: "If x is greater than or equal to zero, it *is* a function." Also take advantage of the opportunity to practice formal mathematical notation. Tell this student, "It's kind of hard to write out what you just told me. Mathematicians have a faster way to write what you said: $x \geq 0$."

Students can also *add items to the range of a relationship* to produce a function. Consider the case of the birthday month relationship in the Function Wall activity. Only January through March are in the range, and every other birth month lacks a mapping. Simply adding a fourth option—"Any other birthday"—creates a function. Anybody who has checked "Other" or "None of the above" on a survey understands that the purpose of that option is to eliminate confusion, and it does so by creating a function.

Students may rely on superficial or misleading indicators such as the so-called vertical line test. Reflect 1.4 poses questions about this method, which students frequently rely on to determine whether a relationship is a function.

Reflect 1.4

What are the limitations of the vertical line test?

Under what circumstances will it produce a false reading?

How does the vertical line test limit or enhance a student's understanding of functions?

Nearly one-third of Even's (1993) 152 prospective mathematics teachers explained functions to students by using the vertical line test. One participant described this test by saying, "[Students] can go through with a ruler or a straightedge and vertically go across the function, looking for places where there are two points" (p. 108). However, some of those same participants struggled to understand the limitations of that test. One participant looked at a circle and contradicted himself: "Like if I was going to have.... Well, uh, a circle is a function, but a circle doesn't pass the line test" (p. 110).

Rules and mnemonics like the vertical line test are only as effective as the conceptual knowledge that underpins them. In some cases, those mnemonics undermine conceptual knowledge. Consider the circle, for instance. A circle *isn't* a function with x mapping to y, because a single x-value in the domain maps to more than one value in the range. The vertical line test works here. But the same graph *does* represent a function if we shift our coordinate system from Cartesian to polar. With polar coordinates, we would create a circle by taking any angle in the domain and assigning it to the same value for the radius. Now we would have a function because every value in the domain (0°, 23°, 198°, 370°, to name just four of them) would map to only one value in the range (the radius, in every case).

If the shift to polar coordinates seems too technical, we can stay on the Cartesian plane. Consider the graph of $x = y^2$. This *is* a function, provided that we define the domain as y and the range as x. But the vertical line test makes no such distinctions: $x = y^2$ will fail the test without any consideration of domain and range. The writers of the draft progression on high school functions for the Common Core State Standards assert, "The vertical line test is problematical, since it makes it difficult to discuss questions such as 'is x a function of y' when presented with a graph of y against x (an important question for students thinking about inverse functions)" (Common Core Standards Writing Team 2012, p. 8).

Consider also the parametric scenario posed by Clement (2001), whose students struggled to answer the question shown in figure 1.6.

> **A caterpillar is crawling around on a piece of graph paper, as shown below. If we wished to determine the creature's location on the paper with respect to time, would this location be a function of time? Why or why not?**

Fig. 1.6. Is a caterpillar's location a function of time? (Clement 2001, p. 745)

Three of five students that Clement interviewed applied the vertical line test directly to the caterpillar's path and decided that they were not looking at a function. They made that determination in spite of the fact that for any time in the domain the caterpillar can only have one position in the range. It cannot be in two places at once.

The vertical line test returns correct results for a limited set of functions. But the vast number of exceptions that must be taught along with the test ("Be careful with inverse functions, polar functions, parametric functions"—and so on) militate against teaching it at all. Instead, urge students to pay deliberate attention to the domain and range of a function. (The domain may not always be the horizontal axis.) Once students have located the domain, they can use their finger to map from an element of the domain to the corresponding element in the range. In some cases, this deliberate movement will rather closely resemble the vertical line test. In the case of inverse functions, though, it will look like a *horizontal* line test. And in the case of polar functions, it will appear to be a *radiating* line test. In each of these cases, students are incorporating the concepts of domain and range into a robust understanding of functions rather than integrating misleading indicators like the vertical line test into a limited understanding.

Developing a Robust Understanding of Domain and Range

A robust understanding of domain and range can be helpful in many jobs, but it is simply indispensable for anyone who designs surveys for the Web. Offer your students the screenshot shown in figure 1.7, and ask them to write a few sentences about what is wrong with it and how they would fix it. Their responses will demonstrate their understanding—or misunderstanding—of functions.

The Concept of a Function

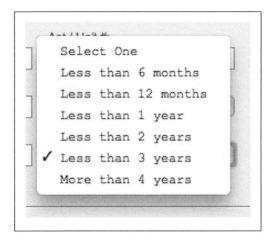

Fig. 1.7. A screenshot of a Web menu of possible survey responses

The designer of the Web menu could have profited at several points from an understanding of functions. Depending on the prompt that preceded this menu, a user might not know which option to select. If the prompt asked, "How long have you been enrolled in this university?" a student who had been enrolled for five months could truthfully select any of those options except "More than 4 years." In this case, we find that the domain (students) maps onto multiple values in the range (length of enrollment).

Furthermore, the domain almost certainly contains a student who simply has no option to select. This particular student does not map to *multiple values* in the range. She maps to no value in the range. Do you see her? She is a student who has been enrolled for 3 years and a few months. She does not have a place in this university's survey. The Web designer has misunderstood the concept of a function in a small but significant way, and the cumulative effect of that small misunderstanding might be an enormous headache months later.

Figure 1.8 shows the Function Finder activity. The activity poses several questions that may advance students' understanding of functions. As your students struggle to grasp the concepts, ask them to describe some hypothetical examples from the domain. They should choose enough examples to make a case that the relationship in question is or is not a function. You should also encourage them to restrict the domain if they think that would be useful.

Function Finder

Question: Which of these relationships are functions?

Goal: Understand the definition of a function.

Decide if each of the relationships below is a function. If it isn't a function, demonstrate how it fails the definition of a function by showing elements of each set that fail it. If it is a function, explain why, and also show several elements of each set.

Facebook user	password	student	hair color	students in our class	planet he/she lives on

state	letters in name	month	days in the month	days in the month	month

date	temperature outside	password	Facebook user	any integer	double that integer

Fig. 1.8. Function Finder activity

Some notes on each of the Function Finder relationships follow:

- **Facebook user → password.** Facebook does not have any mechanism for creating *multiple* passwords, and every Facebook user *must* have a password. Thus, this relationship is a function.

- **Student → hair color.** Whether or not this relationship is a function depends on your domain. If, in your class, nobody has colored or streaked hair, you have a function. If you expand your domain to include the entire school and even *one* student's hair has more than one color, you do not have a function. (You could *make* this a function, of course, by adding a "multicolor" option to the range.)

- **Students in our class → planet the student lives on.** This is certainly a function. Every student maps to exactly one planet. But you might take the opportunity to try to elicit and address a common misconception discussed earlier. Underneath "Earth," write the names of three other planets—perhaps Mars, Jupiter, and Venus. Say to your students, "No one maps to these planets. Can this relationship still be a function?"

- **State → letters in name.** Each state can have only one number of letters in its name (California → 10, for instance). Therefore, this relationship is a function.

- **Month → days in the month.** Some students will leap to the conclusion that this is a function, but others will take their time and recall that February will map to 28 days in roughly three out of every four years but 29 days in leap years. (Students can again exercise some creativity, this time by relabeling the domain "Months except for February" or "Months in years that are not leap years.")

- **Days in the month → month.** Take this opportunity to define the term *inverse*. This relationship inverts the previous relationship, and students should do the same with other relationships on the worksheet after they finish the first task. ("Is the inverse a function?" is a productive question.) The numbers 30 and 31 map to several different months, so this relationship is not a function.

- **Date → temperature outside.** Expect a productive disagreement about this relationship. On the one hand, the temperature may fluctuate wildly over a single day, perhaps from lower in the morning to higher in the afternoon. The relationship is not a function. But if students were to rename this domain "*mean* temperature outside" (which is often the case with historical weather data), they would have a function, since each day has only one mean temperature.

- **Password → Facebook user.** Again, students have an inverse function. In this case, the inverse is *not* a function, because Facebook does not check to make sure a password is unique.

- **Any integer → double that integer.** In considering this scenario, students are inching their way toward pure mathematical functions. Students should see that there is not any way to double 6, for example, and get any other number except 12. The relationship is a function.

Conclusion

This chapter has highlighted possible misconceptions that students may have about the concept of a function. They may hold misconceptions about the single-valued nature of the domain as well as about the multi-valued nature of the range. Students may incorrectly assume that the inverse of a function must also be a function. It is important to confront these misconceptions as early as possible since they will only cause more trouble as students begin to operate on functions, analyze them for covariation, and represent them in different ways.

We encourage you to replicate some of the conversations between Mr. Ramirez and his students. Those conversations are all adapted in one way or another from conversations that we have had with our own students. They demonstrate that it is one thing to define functions from the front of the class, but it is another thing entirely to give students an experience that calls the definition of a function into play and ties it to clarity or confusion in the students' own understanding.

into practice

Chapter 2
Covariance

Essential Understanding 2a
For functions that map real numbers to real numbers, certain patterns of covariation, or patterns in how two variables change together, indicate membership in a particular family of functions and determine the type of formula that the function has.

Essential Understanding 2b
A rate of change describes how one variable quantity changes with respect to another.

Essential Understanding 2c
A function's rate of change is one of the main characteristics that determine what kinds of real-world phenomena the function can model.

Covariation and rate of change are entwined in Big Idea 2 in *Developing Essential Understanding of Functions for Teaching Mathematics in Grades 9-12* (Cooney, Beckmann, and Lloyd 2010). Three associated essential understandings provide a focus on covariation and change with respect to functions—that is, how related quantities vary together. We can classify, predict, and characterize various kinds of relationships by attending to the rate at which one quantity varies with respect to the other. This relationship is called *covariation*. We can all too easily reduce covariation to a simplistic idea of "change." Even if we broaden that definition to include different *kinds* of change—linear, quadratic, exponential, and so on—we can fail to see the difficult mental acts that students must coordinate to understand those changes fully.

Levels of Covariational Reasoning
This chapter offers a hierarchy for assessing a student's understanding of covariation, going far beyond "understands change" or "does not understand change." A precise assessment of a student's understanding of covariation is a necessary but

not sufficient condition for helping the student move higher in that hierarchy. A teacher who has made such an assessment of student understanding must then select and offer tasks and questions that provide opportunities for students to develop conceptual understanding of various contexts of covariance.

Throughout, this chapter emphasizes the importance of directing students' attention to global and local aspects of covariation, as well as to the concrete and abstract aspects of questions. In our own classrooms, we have observed how easy it is to ask students to evaluate a given function by using a table of values. *Local* questions are important for developing understanding of certain aspects of covariation, but they must be interspersed with *global* questions about the context—for example, "What's happening over time? Will the change continue in this direction forever?" A central goal of this chapter is to offer you assistance in recognizing the appropriate time to ask one kind of question rather than another, and the purpose of each.

To begin, let's check back in with Derek and Marta. As you read, try to identify their current understanding of covariation.

> During an afternoon break in the car wash, Derek and Marta are looking at their supplies, trying to figure out whether they will last all day. Throughout the day, Marta has filled in a chart of the number of bottles of soap that their volunteers have used (see fig. 2.1).

When I checked	How many bottles we used since the last time I checked
9:00 a.m.	5
11:00 a.m.	5
Noon.	5
12:30 p.m.	5

Fig. 2.1. Marta's chart of bottles of soap used

"Well, we were certainly consistent," Derek observes.

"If we keep going like this, we are going to have more than enough bottles to finish the car wash," said Marta.

"How do you figure?"

"We are using five bottles every time you check. We have way more than five bottles left."

As Derek and Marta start to pack up and return to the car wash, Mr. Ramirez comes up beside them and looks at Marta's chart.

"You guys may be in more trouble than you think," he says. Seeing their blank looks, he continues, "I think you are misunderstanding something called *covariation* here."

Before reading further, consider your own reaction to Mr. Ramirez's comment, as well as Derek and Marta's possible response. Use the questions in Reflect 2.1 to guide your thinking.

Reflect 2.1

What do you think Derek and Marta are missing in the covariance relationship represented in Marta's chart?

What questions would you ask them if you were their teacher?

How would you assess their understanding of the relationship between time and the number of bottles of soap used?

Operating in real time, Mr. Ramirez must pinpoint what Marta and Derek are missing and move their thinking forward:

> Mr. Ramirez explains that covariation is just a way of talking about how two sets of numbers change in relation to each other. "There are examples of covariation that are counterintuitive and tough to think about and examples that are more obvious. You can first think about covariation being positive or negative," he said. "As your age increases, what happens to your height?"

> "It increases also," said Marta.

> "That's positive covariation. But try another one: If the number of homework problems I assign goes up, what happens to your free time after school?"

> "It goes down," said Derek. "Negative covariation, right?"

> "Right," said Mr. Ramirez. "Now let me show you a couple of places where it's easy to go wrong with covariation."

Carlson and colleagues (2002) report on a study in which calculus students had difficulty forming images of continuously changing rates and made errors in describing inflection points in functions. Students were able to reason about covariational change for contiguous intervals. However, they struggled to represent situations that contained changing rates. On the basis of this study, the authors proposed a "covariation framework," describing five levels of mental actions that support images of covariation. Figure 2.2 shows their framework.

Level	Description
MA1	Coordinating the *value* of one variable with changes in the other
MA2	Coordinating the *direction of change* of one variable with changes in the other variable
MA3	Coordinating the *amount of change* of one variable with changes in the other variable
MA4	Coordinating the *average rate of change* of the function with uniform increments of change in the input variable
MA5	Coordinating the *instantaneous rate of change* of the function with continuous changes in the independent variable for the entire domain of the function

Fig. 2.2. Covariation framework proposed by Carlson and colleagues (2002, p. 357; italics added)

These mental actions are hierarchal and form the basis of the five covariational reasoning levels shown in the chart in figure 2.3. Students at level 1 are able to coordinate the value of one variable with the value of another (MA1) but are likely to struggle with the direction of change (MA2) or the amount of change (MA3). By contrast, students at level 5 are able to coordinate mental actions MA1 through MA5, indicating that they understand instantaneous rate of change, inflection points, and increasing and decreasing areas of a function.

Covariational reasoning level	Supporting mental actions
Level 1: Coordination	MA1
Level 2: Direction	MA1, MA2
Level 3: Quantitative	MA1, MA2, MA3
Level 4: Average rate	MA1, MA2, MA3, MA4
Level 5: Instantaneous rate	MA1, MA2, MA3, MA4, MA5

Fig. 2.3. Covariational reasoning levels with supporting mental actions. Adapted from Carlson and colleagues (2002, p. 358).

Supporting Local and Global Perspectives on Covariation

Students need many experiences to support their development of covariational reasoning and their progress from one level to the next. The three real-world scenarios that follow provide opportunities for you to encourage your students to use both local and global lenses as they reason about covariation. Likewise, the scenarios offer opportunities to assess your students' covariational reasoning before introducing them to more advanced ideas about covariation.

Lake Depth

An example provided by Cooney, Beckmann, and Lloyd (2010, p. 28) highlights a key challenge that students face in understanding covariance. Learners must first understand the pattern of variance for each individual variable and then coordinate those patterns to conceptualize a new relationship between them. Figure 2.4 presents Cooney, Beckmann, and Lloyd's data from the Lake Depth example. Additional columns show the changes in days corresponding to the changes in depth, and Reflect 2.2 invites you to consider for yourself the covariance in the example.

Lake Depth

A park ranger measured the depth of the water in a lake at the same spot over a period of several weeks and recorded the results in a table:

Change in days	Day	Depth	Change in depth
	7	15.29	
7	14	15.43	0.14
14	28	15.57	0.14
7	35	15.71	0.14
7	42	15.85	0.14

Fig. 2.4. Data for the Lake Depth example. From Cooney, Beckmann, and Lloyd (2010, p.28).

Reflect 2.2

How would you describe the covariance of the two variables "day" and "depth of lake" in the Lake Depth example, shown in figure 2.4?

Students may consider only the change in depth and assume that the change is constant across all measures, an assumption that would indicate that they had failed to take into account the *amount of change* (level 3) in the independent variable. Attending to changes in the time intervals for the measurements recorded in the table reveals patterns that might help these students advance to level 3 covariational reasoning. Students should examine how the perturbation in the independent variable affects the covariance relationship. Graphing the data as in figure 2.4 might provide additional insight into the relationship. The graph in figure 2.5 makes the interval between day 14 and day 28 stand out.

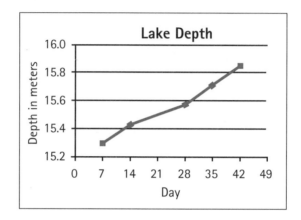

Fig. 2.5. Graph of the lake-depth data from the table in figure 2.4

Another way to analyze students' interactions with, and interpretations of, this representation is to consider their local and global understandings (Leinhardt, Zaslavsky, and Stein 1990) to identify their misconceptions more closely. Students may be able to determine values for specific data points accurately in both the table and the graph, yet not fully connect the table with the graph. Although the students may notice that the rate of change is different in the graph between days 14 and 28 (a global perspective), they may not be able to explain readily why it is different.

If students struggle to understand this relationship in a representation of this form, another representation may prove useful in helping them compare and analyze the variance of both variables more fully. Graphing each variable separately on a single *x*-axis but on different *y*-axes, as in figure 2.6, may be a valuable exercise. In this graph, trend lines highlight the gap in the data for the independent variable. Before students are able to describe the covariance between two variables, they must first be able to describe the variance of each variable accurately and in detail.

Covariance

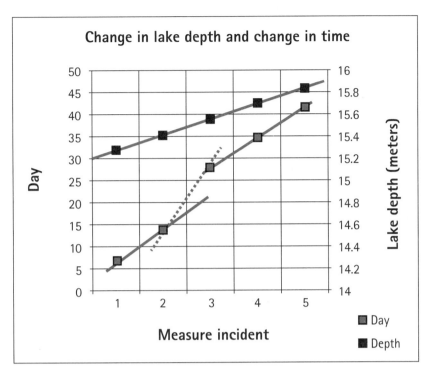

Fig. 2.6. Understanding covariance by examining changes in depth and time

Students who focus on the detailed aspects of the graph, such as values of points and slopes of individual lines, are viewing the representation from a local perspective. Those who focus on the relationship of the lines or the shape of the graph are taking a global view (Leinhardt, Zaslavsky, and Stein, [1990]; see also Chapter 4 of this volume for further discussion of local and global interpretations of graphs). Students may need to move from local to global to local points of view to analyze the meaning of the graph adequately and interpret it accurately with respect to the situation that it models. Questions to ask students to help them consider both local and global perspectives might include the following:

1. In the table of the data for the Lake Depth example (fig. 2.4), what does the gap in days mean about the measurement process? (local)

2. In the graph of the lake-depth data (fig. 2.6), what is the rate of change for each of the lines? (local)

3. What could explain the consistent depth measures in the lake-depth data, considering the gap in those measures? (That is, what happened to the lake?) (global)

4. How does the gap in days in the table of lake-depth data (fig. 2.4) affect the

covariance between the day and the lake depth? (global)

5. Describe the meaning of the dashed line in the graph (fig. 2.6)? (global)

6. What does the overall average rate of change in lake depth tell you about this relationship, and what does it fail to capture? (global)

7. How would you describe the change in lake depth over time? (global)

8. If the rate of change in lake depth remained constant (increase in depth per day), what depth would the park ranger have measured at time 3? At times 4 and 5? (local and global)

To make sense of covariance, students may need many experiences of moving between global and local lenses and analyzing multiple representations of that covariance. In this situation, students have a scenario, or story, along with a table of data and two graphical representations that tell the same story in different ways. You should expect your students to struggle in moving among these representations as they work to make sense of them and the situation that they describe.

Parking Fees

A different scenario involving a piecewise function may help your students develop their understanding of covariance and local and global perspectives on it. Students' work with the Parking Fees scenario may reinforce the notion that change need not be constant or linear, and the scenario may help them connect a change in cost with a changing situation. Figure 2.7 shows the scenario and task for students.

Parking Fees

Unlike many airports that have two different parking rates, one for short-term parking (for example, for passenger pickup) and the other for long-term parking, the local airport adjusts its parking fees automatically according to the time parked, to accommodate different types of use. In the first 12 hours, parking costs $1 for each hour, and after 12 hours, parking costs $10 per day.

  a. Write a piecewise function to describe parking costs at this airport.

  b. Graph your piecewise function.

Fig. 2.7. The Parking Fees scenario and task

Figure 2.8 shows a graph of the piecewise function that models the parking costs. This graphical representation for parking costs is quite different from those that students typically encounter. Yet, it offers unique ways to experience moving between local and global lenses on functions.

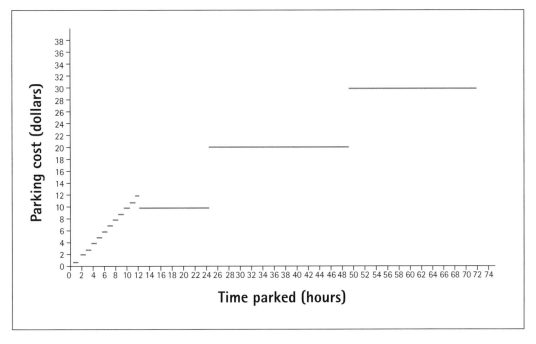

Fig. 2.8. A graph of the airport's parking fee structure

After your students have examined the graph—perhaps after successfully constructing it—ask them the following questions to help them develop their local and global understanding of the situation and its covariance:

1. How do the short line segments in the graph differ from the longer segments? (global)

2. What does the graph indicate that the fee is at 4 hours? (local)

3. Note that the graph seems to overlap at 12 hours. What happens at the parking ticket booth then? (global) What happens at 10 hours? (global and local)

4. What happens from 12 to 24 hours? Why is this line segment a different length from the segments up to 12 hours and the segments after 24 hours? What does this line segment describe about the parking fee? (global and local)

5. How does the fee change with the length of time that a vehicle is parked? (global)

6. What would be the fee after x hours? (global and local)

7. What would be the fee if you parked for 3 days? For 10 days? (local)

Students should be able to describe how the parking fee changes incrementally with time in hours (t) by connecting the short line segments ($t < 12$) with the story. For example, they should explain why the individual short segments are horizontal and why the graph seems to have gaps or steps. Their ability to offer these explanations represents covariational reasoning at level 3 (Carlson et al. 2002), since students are coordinating ideas about how a change in time affects the change in the parking fee. At $t = 12$ hours, the rate of fee increase changes, a realization that students who have level 4 thinking can manage readily. Students could also consider what the fee would be for vehicles parked for times just less than or just greater than 12 hours.

Students should also be able to describe what the short horizontal line segment at $x = 6$ means—both its value and why it is horizontal—as well as what it means about parking fees. Beyond this, they should be able to explain the meaning of the lengths of the line segments and their positions on the graph (global). In particular, they should discuss the meaning of the length of the segment from the 12-hour mark to the 24-hour mark to demonstrate their understanding of the situation fully.

Using both local and global lenses, students should discuss the meaning of covariance for this function. They should be able to describe how the parking cost varies with the number of hours. For example, they should be able to discuss the points at which the covariance changes, when it stays the same (remains invariant), why it stays the same when it does, and what all of these points mean in relation to the story.

Boat Rental

The Boat Rental example presented in figure 2.9 offers students an opportunity to consider and represent a piecewise function in a different way. In this setting, two separate functions describe the rental charges; figure 2.10 shows the graphs of the two functions. Notice that the rates for the functions do not overlap—that is, together they do represent a function with unique outputs.

Covariance

> **Boat Rental**
>
> The community park has a small lake where visitors can rent paddle boats at $1 for every 15 minutes, up to 2 hours. After 2 hours, the rate increases to $3 for 30 minutes.
>
> Write the piecewise function that models this situation, and graph the function.

Fig. 2.9. The Boat Rental example

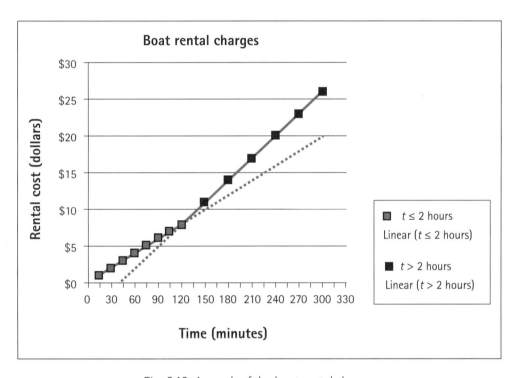

Fig. 2.10. A graph of the boat rental charges

The function that models the Boat Rental example is a different type of piecewise function from the previous one modeling the Parking Fees scenario in that it can be represented with slopes and an intercept. As in the case of the earlier graph in the Lake Depth example (fig. 2.6), students' prerequisite skills should include the ability to discuss slope, intercept, domain, and range. They should demonstrate their understanding of covariance by comparing the meaning of the two lines. Further, they should discuss the two lines and how they are alike and different in terms of slope, their relationship to each other, and their connections with the story (global).

Connecting this graph with the story can lead to interesting strategies for minimizing cost and keeping the boat longer. For example, students could discuss whether two 1-hour trips would be less expensive than one 2-hour trip (local and global). Coordinating the two rates of change requires students to operate at covariational reasoning level 4. However, discussing and explaining the situation at $t = 2$ hours, much like analyzing and interpreting the case of $t = 12$ hours in the Parking Fees scenario, can cause students to address two different rates of change and may set the stage for operating at level 5. Building a table of values to represent this set of fees, or writing an expression for the fee for x minutes (piecewise function), could also develop students' understanding.

A variety of questions can focus students' attention on the big picture and small details, requiring them to examine the Boat Rental example from local and global perspectives. Possible questions to ask include the following:

1. What is the rental charge at 15 minutes? At 16 minutes? At 45 minutes? (local)

2. If you had only $15, how long could you rent a boat? (local)

3. How are the two lines in the graph the same, and how are they different? (global)

4. How would the graph change if the rate change occurred at $t = 1$ hour? (local)

5. How would the graph change if the fee were $2 for 15 minutes? (local)

6. What is the rental fee at 3 hours? (local)

Reflecting on two or three related problems at the same time may lead students to additional generalizations about covariance. Think about and respond to the question in Reflect 2.3, and then pose it to your students.

Reflect 2.3

How is the graph that represents the Boat Rental scenario (fig. 2.10) similar to or different from the graph that represents the Parking Fees setting (fig. 2.8)?

Covariance

When presenting this question to your students, be sure to probe at both global and local levels. Ask how the graphs are similar—for example, in shape (increasing) and structure (cost varies with time, more than one line). Also press students to describe how the graphs are different—for instance, in shape (horizontal lines and slanted lines, number of lines) and structure (time period). This set of questions addresses global qualities, but investigating local qualities is equally important.

Questions with a local focus might probe what a particular line or point means with respect to the story. For example, how does the line that starts at 12 hours in the parking problem compare with the line that starts at 0 hours in the boat rental graph? How are they alike? (Both show cost for a particular time period.) How are they different? (One is increasing, and the other shows a flat rate.) These and similar questions give students opportunities to compare and contrast graphs that they have already encountered and therefore may help them develop deeper understanding of covariance from both global and local perspectives.

Reflect 2.4 connects the Boat Rental problem with the Lake Depth example. Again, respond to the question yourself before posing the same question to your students.

Reflect 2.4

How is the graph of the Boat Rental setting (fig. 2.10) similar to the graph of the Lake Depth example (fig. 2.5)?

From a global perspective, students should recognize that the both the independent and dependent variables are different (day predicting depth vs. hour predicting cost). They should also notice that the line in the Lake Depth problem is not straight. What would such a line mean in the Boat Rental problem? Taking a local perspective, they may investigate how the independent variable changes and how that change affects the dependent variable.

Supporting Reasoning about Quadratic Rates of Change

Linear functions have a single rate of change; that is, the value of the slope remains constant as the independent variable changes. Students can explore the slopes of many lines in a dynamic way by working with interactive tools like GeoGebra (developed by the International GeoGebra Institute and available without charge at http://www.geogebra.org/cms/en). Exploring with the software enables them to develop intuitions about the impact of slope on a line and to explore and compare aspects of a line that vary and those that remain invariant.

Working with quadratic functions in GeoGebra or similar programs allows students to expand on this notion as they observe that the change in the slope (unit rate) changes with the value of the dependent value. Students working with sketches in GeoGebra can use sliders to model linear, quadratic, or cubic functions. The GeoGebra screenshot in figure 2.11 shows a sketch of a quadratic function and a tangent line that captures the slope of the quadratic function at a given point. When the sketch is live in GeoGebra, the points are movable, so students can readily see the changing slope of the function as they move the point along the curve. Although the software provides two points, allowing tangents to be readily compared, students can turn off one by using the check box, so that they can initially focus on the change as they drag one point along the graph of the function.

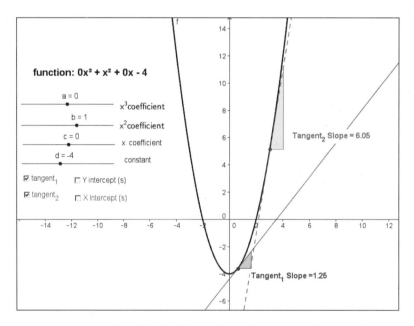

Fig. 2.11. GeoGebra sketch of a quadratic function and tangent lines

Students should understand the relationship between the slope and the tangent. They should discuss the covariance relationship, how it changes continuously, and what the continuous change means about that particular place on the curve. Working with the sketch can help students explain the difference and similarities, using covariational reasoning that falls between level 4 (average rate; slope of a line), and level 5 (instantaneous rate; slope of a tangent at a point). Furthermore, students can compare the slopes of the tangents at different points on the curve and discuss what the differences mean with respect to the shape of the curve and the covariance relationship of the function.

Working in real-world contexts can help students develop their understanding of covariance relationships that involve continuously changing rates of change. Examples involving an object falling from a height, a bottle filling with water, and a ladder slipping down a wall provide useful opportunities to connect a scenario with ideas about covariance and rate of change.

Falling Object

Quadratic functions are used to describe the motion of falling objects. Such a real-world application of a function in the context of a related story can give students additional language for use in describing a covariance relationship. For example, consider how students might explore the case of an object that starts falling at a height of 400 feet—a scenario that is modeled by the function graphed in figure 2.12.

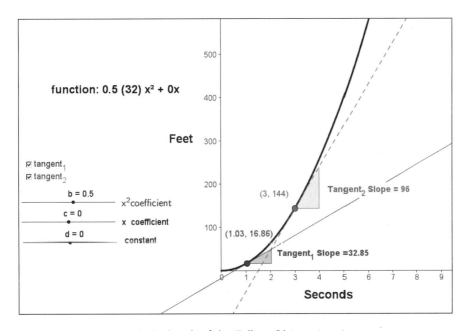

Fig. 2.12. Graph of the Falling Object situation

Posing a few initial questions might provide you with information to use in designing explorations for your students. Use the levels of covariational reasoning developed by Carlson and colleagues (2002) and presented in figure 2.3 as a guide as you ask these questions and interpret your students' responses:

1. What are reasonable values for the domain and range of this function, and why?

2. What does the value 32 represent in the equation of the function?

3. What is the direction of the object? How is this shown in the graph?

4. What is the distance that the object has fallen in 1 second? In 2 seconds?

5. How long does the object take to hit the ground? How is this represented in the graph?

6. What does the shape of the graph indicate about the rate of fall?

Students may not initially realize that the starting height of 400 feet is the upper bound for the domain and zero is the lower bound. The range consists of seconds and is limited to the time that the object takes to hit the ground. The direction of fall is toward the earth, so the graph in figure 2.12 is not a picture, or any sort of image, of the falling object. However, the graph does accurately capture the covariance between the time in seconds from the release of the object and the distance in feet that the object has fallen. In addition to making this graph, students should be encouraged to make a graph of the height of the object with respect to time and compare that graph with this one. Continue posing questions to your students:

7. What does the slope mean at time = 2 seconds? At time = 3 seconds? What will it be at time = 5 seconds?

8. In the graph of distance fallen as a function of time (fig. 2.12), can you find a pattern in the instantaneous slope and the time?

9. What does a slope of zero mean in this situation?

10. What is the maximum rate of change?

11. What does the slope mean in terms of the story?

12. If the object started at 500 feet, how would the graph change?

The answers to these questions require students to move between local and global interpretations of the graph, as suggested by Leinhardt, Zaslavsky, and Stein (1990). Quadratic equations help students gain an understanding of the difference between slope and instantaneous slope—the advance described by Carlson and colleagues (2002) as marking the difference between level 4 and level 5 covariational reasoning and guiding students to adeeper understanding of these functions.

Students can use any graphing tool to create the functions shown in figures 2.11 and 2.12, and they can employ the dynamic features of the software to study how the functions change as their parameters are changed. As noted earlier, the GeoGebra tools used to create the sketches shown in these figures are available online and could be a useful resource for your students.

The next two examples involve functions that describe increasingly complicated covariant relationships. The stories for these involve a bottle that is filling with water at a constant rate and a ladder that is sliding down a wall at an increasing rate.

Filling Bottle

Carlson and colleagues (2002) explored students' thinking about covariance by using the Filling Bottle problem, shown in figure 2.13. Students were told that the bottle was filling with water at a constant rate, and their task was to sketch a graph of the height of the water as a function of the amount of water in the bottle.

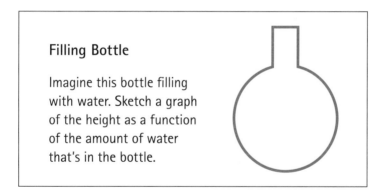

Fig. 2.13. The Filling Bottle problem. From Carlson (2002, p. 358).

Students whose covariational reasoning is at level 1 (coordination) will be able to identify the independent and dependent variables and will note that as one variable changes (the volume of water), the other will also change (the height of the water in the bottle). Students who have reached level 2 (direction) will also be able to express the idea that the direction of change for both variables in this example is the same. Students whose covariational reasoning has progressed to level 3 (quantitative coordination) will be able to coordinate the amount of change in the height of the water with the change in its volume. They may exhibit understanding at this level by marking incremental heights on the bottle in steps that decrease to the widest part of the bottle and increase after that point (see fig. 2.14), or by plotting points on a graph.

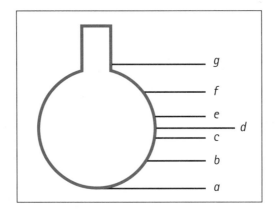

Fig. 2.14. Level 3 analysis of height and volume in the Filling Bottle problem

Students who have reached level 4 (average rate) will be able to coordinate the average rate of change of the height with respect to the volume. Students at this level may construct small contiguous line segments with different slopes that reflect the rates of change at particular heights to account for the shape of the bottle (for example, at points *a–g* in figure 2.14).

Finally, students who are reasoning at level 5 (instantaneous rate) will be able to construct a smooth curve that reflects the varying rate of change of the height of the water in the bottle. In this instance, the curve would first be concave down, then would appear to be linear for a short way (the visual impact of the inflection point), and then would be concave up (see fig. 2.15).

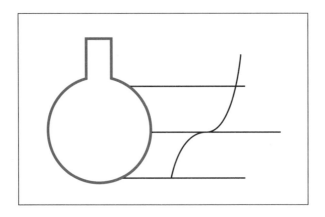

Fig. 2.15. Level 5 analysis of height and volume in the Filling Bottle problem

Students should be able to determine where the height of the water will rise slowest (at point *d*) and fastest (at point *g*), and indicate how the water will rise at points between and why it will rise that way. They should be able to connect the rate of change in the height of the water with the shape of the bottle, as in figure 2.14. For example, students should realize since the water is filling the bottle at a constant rate, the change in height will slow as the container widens and speed up as it narrows. They may not realize that the change is nonlinear. To be able to coordinate the independent variable (amount of water) with the dependent variable (height), students must integrate both global and local views. By presenting your students with scenarios involving containers of different shapes, you can build their experience with this type of rate of change.

Sliding Ladder

Monk (1992) used a calculus problem to study students' understanding of functions and their ability to distinguish the graph from the path of the object being modeled. Monk drew this problem from a 1989 Addison Wesley calculus textbook; the original problem statement appears in figure 2.16.

Sliding Ladder

A 14-foot ladder is leaning against a house when its base starts to slide away. By the time the base is 12 feet from the house, the base is moving at a rate of 5 ft/sec. How fast is the top of the ladder sliding down the wall then?

Fig. 2.16. The Sliding Ladder problem.
Used by Monk (1992, p. 181) from a 1989 Addison Wesley calculus textbook.

In Monk's study, students were provided with a physical scale model (1 in. = 1 ft.) to help them visualize the problem. The model gave the ladder in its initial position, which was almost vertical against the wall, and then in two subsequent positions, in which (1) the base of the ladder had moved a certain distance from the wall, and (2) the base of the ladder had moved another equivalent distance from the wall. Students were then asked to describe the distance that the top of the ladder moved down the wall: was it the same distance that the base moved, a distance greater than the base moved, or a distance less than it moved? Pause to respond to the questions in Reflect 2.5 before reading further.

Putting Essential Understanding of Functions into Practice in Grades 9–12

Reflect 2.5

Monk (1992) asked students whether the distance that the top of the ladder moved down the wall was the same as, greater than, or less than the distance that the base moved.

How would you answer Monk's question?

How do you think your students would respond to it?

Although Monk provided students with physical models to explore this phenomenon, interactive geometry systems can offer your students a more structured method. Figure 2.17 shows sketches created in GeoGebra to represent the initial position and two subsequent positions of the ladder in its slide; here the length of the ladder is 18 feet rather than 14 feet, as in the problem used by Monk. GeoGebra allows users to move the base of the ladder 1 foot at a time (other increments are possible), and the software can readily record the movement of the top of the ladder. In fact, GeoGebra offers a spreadsheet view that students can use to capture the data automatically as the ladder moves.

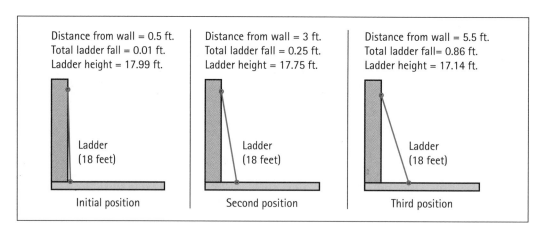

Fig. 2.17. Images from GeoGebra modeling an 18-foot ladder at initial, second, and third stages of sliding down the wall

Some initial questions may help guide students in understanding this function:

1. Will the top or the bottom of the ladder move faster?

2. Will the top and the bottom of the ladder move at the same rate? When?

3. If the bottom of the ladder moves one foot, how far will the top slide down the wall?

4. Will the bottom of the ladder always move faster than the top?

5. How can you compare the change in the movement of the top of the ladder with the change in the movement of the bottom of the ladder?

Before investigating this situation, students should be encouraged to predict the behavior of the movement of the top and the bottom of the ladder. Once they have had an opportunity to discuss the movement of the ladder, they should explore the motion of the ladder and examine how the top and bottom distances vary in a physical model. (Monk used a cardboard ladder and wall.) As they study the movement of the ladder, students should be encouraged to take measurements and construct a table of values such as that in figure 2.18 and plot the values in graphs, as shown in figures 2.19 and 2.20. Note that in the table in figure 2.18—

- Base = horizontal distance (ft.) of the bottom of the ladder from the wall;

- Fall = vertical distance (ft.) that the ladder has fallen; and

- Height = vertical distance (ft.) of the top of the ladder from the ground as it falls.

Base	Fall	Height	Base	Fall	Height	Base	Fall	Height
1	0.03	17.97	7	1.42	16.58	13	5.55	12.45
2	0.11	17.89	8	1.88	16.12	14	6.69	11.31
3	0.25	17.75	9	2.41	15.59	15	8.05	9.95
4	0.45	17.55	10	3.03	14.97	16	9.75	8.25
5	0.71	17.29	11	3.75	14.25	17	12.08	5.92
6	1.03	16.97	12	4.58	13.42	17.99	17.40	0.60

Fig. 2.18. Data (feet) generated for an 18-foot ladder sliding down a wall

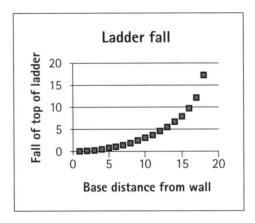

Fig. 2.19. A graph showing the amount of the ladder's fall against the distance of its base from the wall; distances in feet

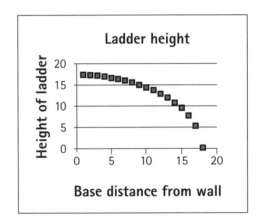

Fig. 2.20. A graph showing the height of the ladder against the distance of its base from the wall; distances in feet

In Monk's study, students were asked a second question, which involved predicting the fall of the ladder, given repeated and continuous movement of the base of the ladder away from the wall. The graph in figure 2.19 compares the change in the ladder's base with the change in the ladder's height (that is, the amount of its fall). This graph plots the values in the table in figure 2.18 for the first dependent variable. Your students' interpretations of the second dependent variable in the table can provide you with an additional assessment of their understanding of this relationship. In this interpretation of the situation, the height of the ladder is measured each time the base is moved one foot from the wall; the graph in figure 2.20 plots the values for this variable. Some students may be able to visualize height better than the fall of the ladder.

By focusing locally on the movement of the ladder, students may attend to individual increments or changes in each ladder movement. A global focus may help students consider the overall shape of the graph or the change in slope. This problem provides opportunities for students to demonstrate understanding at every level of Carlson and colleagues' covariational reasoning framework. The rates of change of the bottom and the top of the ladder may be perplexing to students at first, since the top and the bottom are bound together by a physical object—the ladder. Yet, they move in different directions and at different rates. Understanding average rate and instantaneous rate will be important for students as they sort out the covariation in this function. The next section examines how students might disentangle this covariation in responding to the third question that Monk posed to students.

Interpreting the Covariance of Two Variables

It is important for students to have experiences in interpreting the covariance of two variables to help them understand that a function expresses how two variables covary. Teachers may need to assess their students' understanding of the concept of a variable before assessing their understanding of the covariance between two variables. The idea of variable itself is difficult for students to understand (Philipp 1992; Shoenfeld and Arcavi 1988; Akgün and Özdemir 2006).

Küchemann (1978) organized student misconceptions of variables into two Piagetian subscales (concrete operations and formal operations) with three levels in each. The misconceptions that fall into formal operations (levels 4–6) are of particular interest in helping students understand the covariance of two variables. Understanding a variable as a specific unknown (level 4) means knowing that a variable that is represented by a letter stands for a number that can be operated on without being evaluated. Understanding a variable as a generalized number (level 5) means recognizing that a variable can stand for a series of numbers, a list, or a matrix. Understanding variable as a functional relationship (level 6) means understanding that two variables change together, or are coordinated.

The Sliding Ladder problem addresses understanding at level 6 of Küchemann's hierarchy, as well as the ability to translate between representations of functions (Monk 1992; Russell, O'Dwyer, and Miranda 2009; Clement 1989). Once students are able to demonstrate understanding of the various representations of the Sliding Ladder problem (see figs. 2.17–2.20), one more step remains that offers a further opportunity for assessing their understanding: they must demonstrate the ability to interpret rate of change.

Initial questions may prove useful for exploring students' understanding of rate of change, or in this case, the speed, of the ladder's fall in the problem posed by Monk (1992). These questions may include the following:

1. Is the rate of fall of the top of the ladder increasing, decreasing, or constant?

2. When is it falling fastest?

3. How fast is it falling at that point?

4. When is it falling slowest?

5. What is its rate of fall at that point?

Before students try to interpret or develop a model to represent the speed of the ladder's fall, it might be useful for them to think about the possibilities for the two variables to covary. Returning to GeoGebra or a physical model is a good strategy for exploring these relationships (see fig. 2.17). Once students have addressed these questions in connection with the model, useful next steps are analyzing the table and the corresponding graph with respect to the story of the falling ladder.

The third question that Monk posed in his study asked students to describe the rate of fall in relation to time, given that the base of the ladder is moving at a steady rate away from the wall. Figure 2.21 shows a set of data and a graph depicting a continuous movement of an 18-foot ladder in a situation where the base is moving one foot from the wall every second (independent variable), and the top of the ladder moves a varying amount.

First, students should be able to describe the rate of change, or speed, of the falling ladder at each interval. For example, at time = 4 seconds, in one second the ladder fell from 0.25 feet to 0.45 feet (as shown in the table in fig. 2.18), or at a rate of 0.20 feet per second (as shown in the table in fig. 2.21), or 2.4 inches per second. How is this rate expressed in the table versus the graph? Moreover, how does this interval change relate to the total distance that the ladder fell or the height of the ladder (see the table in fig. 2.18)? Next, contrast this rate of change with that later in the sequence; say, at time = 10 seconds. In this case, the ladder moved from 2.41 feet from the top to 3.03 feet, or a change of 0.62 feet, which corresponds to a rate of 0.62 feet per second, or about 7.4 inches per second. Exploring these rates of change while answering the questions posed above can serve to build students' understanding of the covariation of the two variables time and distance in the Sliding Ladder problem.

Monk (1992) found that many students thought that since the base of the ladder was moving at a constant rate of speed, the top of the ladder would also move at a constant rate of speed. Other students expressed a vague notion that the base and top of the ladder would move at the same speed when the ladder was at a 45° angle.

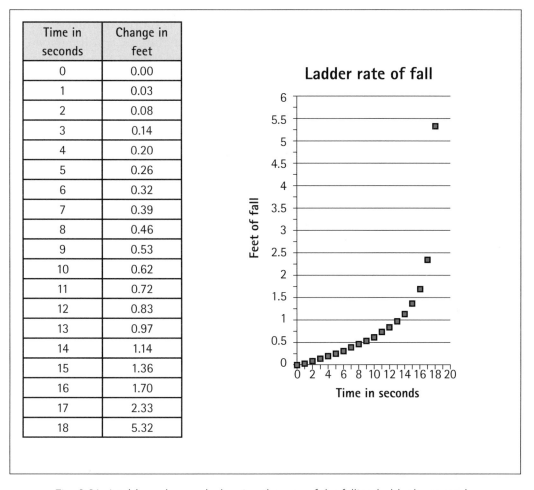

Fig. 2.21. A table and a graph showing the rate of the falling ladder by second

Gathering, plotting, and analyzing the points of the graph of the ladder's movement can provide students with a variety of rich experiences that may move them from a static to a dynamic understanding of graphical representations of functions.

Conclusion

The contribution of Carlson and colleagues' (2002) covariational reasoning framework to our understanding of how students understand covariation cannot be overstated. Their hierarchy provides not only a useful diagnostic tool for assessing students' understanding of covariation but also a roadmap that points them in a productive direction. This chapter has attempted to flesh out Carlson and colleagues' framework of covariational understanding with pedagogical suggestions and examples. If your students struggle with the transition from level 4 reasoning (average rate of change) to level 5 reasoning (instantaneous rate of change), they may profit from the questions suggested here about a falling ladder. If they demonstrate difficulty even with covariational reasoning at level 1, struggling to coordinate two variables, you may find it useful to ask them to explore the Parking Fees or Boat Rental scenario. Once they can successfully describe the direction of the covariation in those situations (constant at times, increasing linearly at other times), thus exhibiting level 2 reasoning, asking them to draw even a staggered, incremental graph of Carlson's bottle may move them toward an initial quantitative understanding of covariation, consistent with level 3 covariational reasoning.

But you should not assume that the only prerequisite for student learning is the right question posed by you at the right time. You can help in other ways, too. In particular, you can draw your students' focus repeatedly to both the global and local aspects of a context. If students struggle to begin their ascent up Carlson's hierarchy, global questions may help—for example, "What do you think happens to the height of the water in the bottle if we add more water?" Such questions may seem obvious and intuitive to mathematics teachers, but for students struggling to make the transition from level 1 to level 2 covariational reasoning, they are ideal. As students progress through the hierarchy, ask questions that focus locally on precise values: "How high off the ground is the ladder after 10 seconds? How about 11 seconds? OK, now pick another one-second interval in the future. Will the change be the same?"

As we have seen, it is also helpful for students to coordinate the abstract and concrete representations of these tasks. It is possible for students to spend too much time immersed in a single graphical or tabular representation of a context. To prevent the condition that Hayakawa calls "dead-level abstracting" (1952), in which a student is unable to identify what numbers and expressions mean as concrete objects, ask repeatedly, "What does the slope [or value, or change] mean in terms of the story?"

Clearly, a fluid understanding of covariation, one that moves swiftly up Carlson's hierarchy, does not come cheaply. It requires acts of mental coordination on the part of the student and acts of pedagogical coordination on the part of the teacher.

Chapter 3
Combining and Transforming Functions

Essential Understanding 4a
Functions that have the same domain and that map to the real numbers can be added, subtracted, multiplied, or divided (which may change the domain).

Essential Understanding 4b
Under appropriate conditions, functions can be composed.

Essential Understanding 4c
For functions that map the real numbers to the real numbers, composing a function with "shifting" or "scaling" functions changes the formula and graph of the function in readily predictable ways.

Under appropriate conditions, functions can be combined, composed, and transformed, and their subsequent behaviors can be predicted. This chapter focuses on essential understandings that capture these ideas, presented as Essential Understandings 4a, 4b, and 4c in *Developing Essential Understanding of Functions for Teaching Mathematics in Grades 9–12* (Cooney, Beckmann, and Lloyd 2010). Supporting these insights in the classroom represents a significant leap from the work discussed in Chapters 1 and 2 of helping students distinguish functions from relationships and developing their ability to analyze functions for covariation.

Once your students have progressed to this point, they can begin to understand how to use the functions themselves as values for other functions. Doing so represents a steep climb from their previous vantage point on functions, and misconceptions may arise as they start combining, composing, and transforming functions.

Although you need to anticipate and attend to these varied misconceptions, none are quite as significant as those related to the domain and range of the functions that result from these operations (Dotson 2009; Gur and Barak 2007). In truth, these operations can be performed with procedural skill alone. Students who know how to combine like terms, multiply monomials, and simplify rational expressions will produce passable answers when they apply those skills to functions.

However, your students will reveal their *conceptual* understanding of these operations when they can explain how the domain and range of the resulting function differ from the domain and range of either of the two original functions (Bayazit 2011; Vinner 1983). Perhaps there are values of x that they can evaluate with $f(x)$ but that do not make sense with $f(g(x))$. Individually, $f(x)$ and $g(x)$ might both return every real value in their range, but once the students divide those functions, there may be certain values that they can no longer evaluate. For example,

$$f(x) = \frac{x-3}{x-1} \text{ and } g(x) = \frac{1}{x-3}$$

are defined for all integers except $x = 1$ for $f(x)$ and $x = 3$ for $g(x)$. Their product, by contrast, is defined for all integers except $x = 1$. Likewise, $h(x) = \sqrt{x}$ and $k(x) = x\sqrt{x}$ are defined for all real numbers $x \geq 0$, yet their product is defined for all real numbers. Similarly, $f(x) = x$ and $g(x) = x + 1$ are defined for all real numbers, but their quotient must exclude $x = -1$.

Your students' responses to questions about the domains and ranges in situations such as these will illuminate their conceptual understanding and misunderstanding of the combination, composition, and transformation of functions. To help you attend to their misconceptions, this chapter offers sample tasks and questions and simulated student-teacher dialogue. To begin, consider the situation presented in Reflect 3.1.

Reflect 3.1

Suppose that you are talking with a student who says that he was looking at the ripples formed when he threw a rock into a lake. He says, "The ripples looked like concentric circles with radii that were growing by about one inch every second. I know there is a function that gives me the area of a circular ripple in terms of its radius. Is there a function that tells me the area of a circular ripple in terms of time?"

How would you respond to the student?

Combining Functions

Students frequently struggle as they attempt to understand and identify the domains and ranges of functions, particularly when functions are combined. Multiple experiences can sharpen their insight and decrease their confusion. Presenting examples of addition, subtraction, multiplication, and division of functions in the context of stories and concrete representations can be invaluable. Students should be able to explain the meaning of the function in relation to the modeled situation, both before and after the combination. Ask your students to compare the domain and range of each function with those of the resulting combined function. Such experiences can help students connect the domain and range of functions not only with the symbolic representations of functions, but also with the stories that they model. Students should realize why understanding the domain and range is important and how an inappropriate domain or range can lead to nonsensical results.

For example, using Heron's formula to find the area of triangle ABC shown in figure 3.1 leads to a function that gives the area in terms of b, the length of side AC:

$$f(b) = \sqrt{s(s-6)(s-10)(s-b)},$$

where s is the triangle's semiperimeter, and

$$s = \frac{16+b}{2}.$$

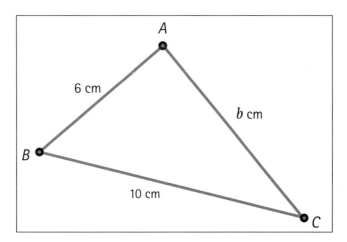

Fig. 3.1. Finding the area of triangle ABC by using Heron's formula requires restricting the domain of the resulting function.

If $b = 12$, then $s = 14$, and $f(12) = \sqrt{14(8)(4)(2)} = 8\sqrt{14}$ square centimeters. However, if $b = 18$, then $s = 17$, and $f(18) = \sqrt{17(11)(7)(-1)} = i\sqrt{1309}$ square centimeters. This result is obviously a nonsensical answer that came about because of a student's incomplete understanding of a function's domain and range, since according to the triangle inequality theorem, the sum of the lengths of two sides of a triangle must be greater than the length of the third side, or, in this case, $4 < b < 16$.

Addition of functions

Extending the car-wash scenario introduced in Chapter 1 can provide an opportunity for students to add functions while giving special attention to domain and range. The expanded scenario includes a new development: Derek and Marta are now working to advertise the event:

> Derek and Marta are creating posters to advertise their service club's car wash. Derek is able to make 2 posters each hour, but once every 2 hours he makes a flawed one, which he discards. Marta makes 3 posters each hour, but she discards 1 poster every 3 hours.

This extension of the original car-wash scenario provides a context in which your students might perform a sequence of tasks:

- Write a function that models the number of posters that each student makes per hour, where x represents the number of hours.

- Combine the functions to get an overall rate of poster making.

- Graph $f(x)$, $g(x)$, and $f(x) + g(x)$, and describe how the three functions are the same and different.

The following functions model Derek's and Marta's numbers of posters:

$$\text{Derek's number of posters:} \quad f(x) = 2x - \frac{1}{2}$$

$$\text{Marta's number of posters:} \quad g(x) = 3x - \frac{1}{3}$$

Combining the functions yields an overall rate of poster making for the two students:

$$f(x) + g(x) = 2x - \frac{1}{2} + 3x - \frac{1}{3} = 5x - \frac{5}{6}$$

Inspecting a graph of $f(x)$, $g(x)$, and $f(x) + g(x)$ can help students describe how these functions are the same or different. Figure 3.2 shows such a graph.

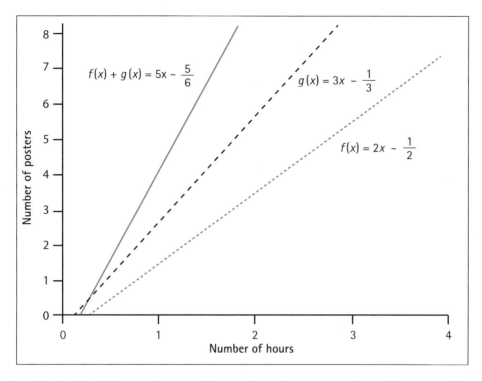

Fig. 3.2. Graphs of the two functions for Derek's and Marta's numbers of posters and the function that represents their sum

Some guiding questions may help the students explore the domain and range of these functions, connect the graph with the symbolic representations, and interpret both graph and symbolic expressions in relation to the story. A few such questions follow:

1. What is the meaning of the slope of $f(x)$, $g(x)$, and $f(x) + g(x)$ in terms of the graph and the story?

2. What does the intercept tell you about the making of posters?

3. How would $f(x)$ change if Derek had to discard one poster each hour?

Students' responses to this set of guiding questions can help you determine their understanding of slope and intercept and gauge how well they connect the symbolic, graphical, and narrative representations.

Continue to question your students, at this point homing in on their understanding of the domain and range:

1. What is the domain of $f(x)$, $g(x)$, and $f(x) + g(x)$? How do you know?

2. What would be the meaning of an x-value greater than 60 for $f(x)$, $g(x)$, and $f(x) + g(x)$?

3. Compare the ranges of $f(x)$, $g(x)$, and $f(x) + g(x)$, and explain your thinking.

4. How is the range of $f(x) + g(x)$ related to those of $f(x)$ and $g(x)$?

These questions may help you identify students who are struggling with aspects of the domain and range of functions. After students have determined the specific domain and range of each function to find meaning at a local level, describing these values in terms of the graph and the story can help them find meaning at a global level. Reflect 3.2 directs your attention more closely to the guiding questions.

Reflect 3.2

How would you expect your students to respond to the guiding questions in the two preceding groups, and how would you explain the answers to your students?

Subtraction of functions

By developing another aspect of the car-wash scenario, we can provide a setting in which students can explore the subtraction of functions while considering questions related to domain and range.

> Derek and Marta are trying to predict the profit that the car wash might yield. They determine that they need to consider the upfront costs of posters that they are planning to create to advertise the car wash and supplies that they will need for washing cars, such as soap, towels, and buckets, regardless of the number of customers. What other costs might they encounter? They also know that some supplies, such as soap, window cleaner, sponges, and paper towels, will be used up by the car wash, and the quantities that they need will be determined by the number of customers. They estimate the average cost per car to be $0.50, and they expect the initial costs to be

$220. The service club has received a $50 donation from the local chamber of commerce to support the car wash, and Marta and Derek take that gift into consideration in deciding that the fee for washing each car will be $10.

This scenario presents students with a rich context for modeling with and subtracting functions, thus obtaining a function that they can use to predict a real-world outcome: profit. Given that Derek and Marta estimate the average cost per car to be $0.50 and expect the initial costs to total $220, your students can write a function—say, $c(x)$—to represent the cost of the car wash, with x representing the number of customers:

$$c(x) = \frac{1}{2}x + 220$$

Next, given that the fee to be charged for washing a car is $10 and the service club has received a $50 donation from the local chamber of commerce, the students can write a function—let's say $r(x)$—to express the receipts from the car wash:

$$r(x) = 10x + 50$$

They can then use these two functions to write a function that models the profit of the car wash, $p(x)$, in terms of the number of customers:

$$p(x) = r(x) - c(x) = 9.5x - 170$$

Figure 3.3 shows the graphs of $c(x)$, $r(x)$, and $p(x)$.

To probe your students' understanding of this example, you can modify some of the questions suggested earlier about slope, domain, and range and pose them, along with additional questions that explore the meaning of the graphs shown in figure 3.3 with respect to this story. A few such questions follow:

- What is the meaning of the y-intercept of the cost function? Of the receipts function? Of the profit function?

- How does the y-intercept of the profit function relate to the cost and receipts functions?

- What are the coordinate values and meaning of the intersection of the cost and receipts functions? What does this x-value mean with respect to the profit function?

Putting Essential Understanding of Functions into Practice in Grades 9–12

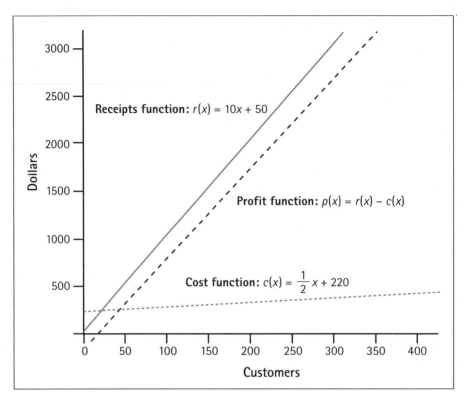

Fig. 3.3. Graphs of the two functions for cost and receipts and their combination (profit) obtained by subtraction

Reflect 3.3 asks you to think about the approach that you would be likely to take with these questions in your classroom.

> **Reflect 3.3**
>
> **You can pose the questions listed above to your students to assess their understanding of the cost, receipts, and profit functions for the car wash.**
>
> **How would you address these questions with them?**

In this case, the cost function is expressed in positive values, so students must subtract it from the receipts function to get the difference—the profit function. In this context, students can explore the meaning of the y-intercept. For example, you should guide the students in explaining the meaning of the *y*-intercept of the profit function. Determining the value of this intercept is an initial step, but explaining the meaning of that value is more important. Students may respond that it suggests that if the car wash draws no customers, the service club will suffer a loss. Encourage your students to compare the slopes of the cost and receipts functions, $c(x)$ and $r(x)$, and express the meaning of these slopes in terms of the story. Pose the following questions:

- How do these slopes manifest themselves in the profit function?

- What does the slope of the profit function mean in terms of the story?

Multiplication of functions

Leaving the car-wash context, students can explore a very different real-world example that is readily modeled by three functions, one of which is the product of the other two. A story about the steady enlargement of a rectangular field in a jungle lends itself to a meaningful exploration of the multiplication of functions.

> Workers begin clearing a rectangular field in a dense jungle. In the first year, they clear a field 3 meters long by 2 meters wide. Each year, they expand the field by 1.5 meters in length and 1 meter in width.

You might launch the investigation by asking your students to consider the area of the field after 5 years. The students can write a function in terms of years to model the changing length of the field, and another function, also in terms of years, to model its changing width. They can combine these functions to represent the changing area of the field in terms of years. Functions for the length (l), width (w), and area (A) follow, with x representing the number of years.

$$l(x) = 1.5x + 3$$

$$w(x) = x + 2$$

$$A(x) = l(x) \times w(x) = (l \times w)(x) = (1.5x + 3)(x + 2) = 1.5x^2 + 6x + 6$$

Graphing these functions is an important next step in solidifying and developing students' understanding. Figure 3.4 shows the graphs.

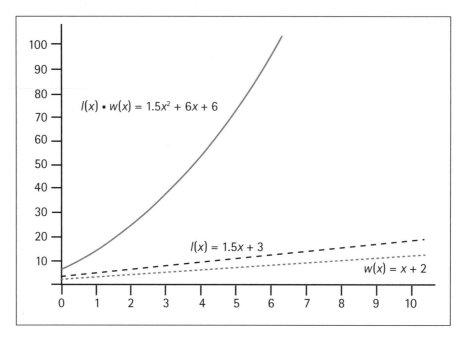

Fig. 3.4. Graphs of the functions for the field's length and its width, in meters, as well as the function for its area, in square meters, produced by combining the first two functions by multiplication, with x representing the number of years

Multiplication of functions provides another opportunity for students to explore graphs at local and global levels. To "think locally," students might evaluate $l(x)$ and $w(x)$ at specific values of x, such as $x = 5$, and multiply those results to compare the product with the output of the functions' product. Be sure to emphasize understanding the meaning of x, $f(x)$, and $g(x)$ in your students' explorations. A global perspective might take in the shapes of the graphs and recognize why the two initial functions are represented by straight lines, although the graph of their product is a curve. Local and global lenses provide students with tools to help them understand not only the behavior of the functions but also their relationships with one another.

As this chapter has consistently emphasized, students should also explore the domain and range of each of the functions and be able to see the impact of the ranges of the linear functions on the range of the quadratic function. They should be able to indicate that the story itself may impose limitations on the domains of the linear functions—the rectangular field cannot expand forever. Useful questions to ask students about the functions and their graphs in relation to the story include the following:

Combining and Transforming Functions

1. What is the meaning of the y-intercepts of the linear functions and the quadratic function? How do these intercepts relate to the story?

2. What do the different slopes of the length and width functions mean about the changing shape of the field?

3. What is the y-value at a given x-value for each of the functions, and what is the meaning of the y-values in terms of the changing field? What does this x-value mean with respect to the area of the field?

Division of functions

Returning to the car-wash scenario, we can develop a new opportunity for students to explore combining functions while examining the impact on domain and range. Extending the situation with Derek and Marta once again provides a context in which students can find real-world meaning in combining functions. This time students divide functions while building on their earlier work of subtracting functions. Previously, they analyzed the club's profit as the difference between its receipts and its costs for the car wash, writing the profit function as $p(x) = r(x) - c(x)$, where x represents the number of customers. Now, by dividing this function, they can consider the average profit per customer.

Students can create a profit-per-customer function by reasoning that

$$\frac{r(x) - c(x)}{x}$$

gives the average profit per customer, for any given number of customers x. They can then write this as a function, $ap(x)$, which returns the average profit per customer:

$$\text{Average profit: } ap(x) = \frac{r(x) - c(x)}{x}$$

Using their earlier work gives

$$\frac{(10x + 50) - \frac{1}{2}x + 220}{x} = \frac{9.5x - 170}{x}.$$

Putting Essential Understanding of Functions into Practice in Grades 9–12

Students should make and analyze graphs of the cost, profit, and average profit functions such as those that appear in figure 3.5, where the average profit function is shown separately in (b), enlarged for visibility, as a "zoom in" representation.

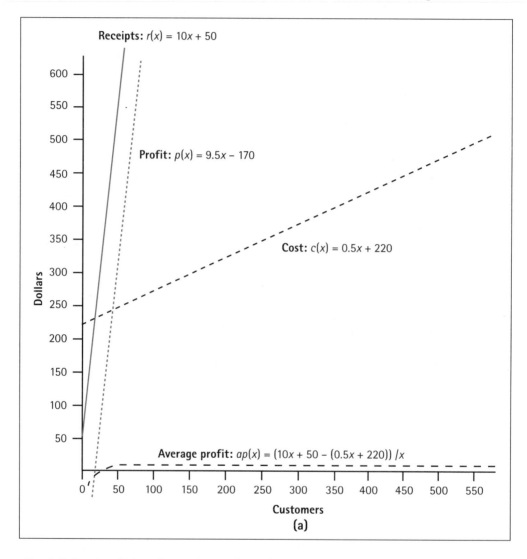

Fig. 3.5. Graphs of (a) profit, receipts and cost functions and (b) the average profit function obtained by division (by the number of customers, x), shown greatly enlarged

Combining and Transforming Functions

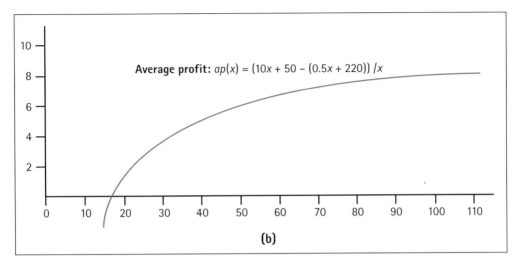

Fig. 3.5. Continued

To help your students interpret the graphs in relation to the story, you can ask questions such as the following:

1. What is the average profit for 100 customers? For 300 customers?

2. What is the result when two polynomial functions are divided? (What are the restrictions on *denominator values*?)

3. Why does the average profit hover around $7.50? What does this value represent?

4. Why is the average profit function so different from the profit and cost functions?

5. Why do the profit and receipts functions look so similar to each other?

The reason why the average profit hovers around $7.50 is important. Reflect 3.4 asks you to pause to consider how you might explain its significance.

Reflect 3.4

How would you explain to your students why the average profit hovers around $7.50?

Students may struggle with the average profit function (see the zoom-in box in fig. 3.5b), since its behavior is different from many that they have encountered. Students should describe what the profit function represents when it is below the x-axis and what the x-intercept means. Students may find that building tables of values for all three functions helps them understand connections with the story (local) and enables them to offer a better description of how these functions are alike and how they are different (global).

A summative activity in combining functions

The preceding discussion has offered some productive tasks involving the combination of functions that are expressed symbolically. Symbolic algebra helps to make the result of combining or composing two functions explicit, but asking students to work exclusively with the graphical representations of functions may sometimes be helpful. The task in figure 3.6 takes the tool of symbolic manipulation out of their hands and asks them to examine two functions shown only in graphs and make broader, qualitative judgments about combinations of the functions.

As students try to locate the x-intercepts of the graph of $f(x) - g(x)$, it will be useful to point out that $f(x) - g(x)$ will touch the x-axis only if $f(x)$ and $g(x)$ have the same value. Likewise, $f(x) + g(x)$ will touch the x-axis if $f(x)$ and $g(x)$ have *opposite* values. Further, $f(x) \times g(x)$ will be zero when either $f(x)$ or $g(x)$ has a value of zero, and $f(x) \div g(x)$ will be zero only when $f(x) = 0$. (The combination by division also leads to a domain restriction where $g(x) = 0$; that is, it restricts the domain so that $g(x)$ cannot be zero.)

Clearly, this kind of activity is useful for reviewing the zero product property, the multiplicative identity property, and the additive identity property. Asking students to *graph* the combined functions is also a worthwhile group task, as individual students will have different observations to contribute. For example, one student may understand that the order of the function will not change for the sum and difference combinations, another may have located the y-intercept, and still others may have pinpointed the x-intercepts. Bringing the students together to share their observations will produce more accurate graphs.

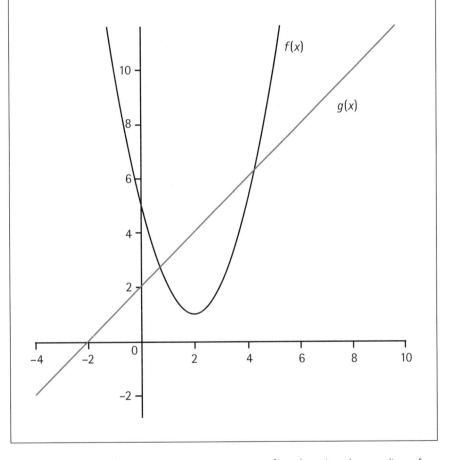

Given the two functions whose graphs are shown, tell everything you can about the following, being as specific as possible:

- $f(x) + g(x)$
- $f(x) - g(x)$
- $f(x) \times g(x)$
- $f(x) \div g(x)$
- The x- and y-intercepts in the combined functions
- The shapes of the graphs of the combined function

Sketch each of the combined functions.

Fig. 3.6. A task for a summative assessment of students' understanding of function combinations

You should feel free to reintroduce the symbolic forms of these functions after your students have exhausted their graphical interpretations. The explicit forms of the two functions follow:

$$f(x) = x^2 - 4x + 5$$

$$g(x) = x + 2$$

At this point, your students will have sketched each of the four combinations. You can use a graphing tool to graph each one precisely so that students can check their work.

Composing Functions

The car-wash scenario offers possibilities for introducing students in a natural way to the potentially challenging idea that functions can be composed. Students can find meaning in the idea of composing functions in the following vignette:

> The car wash is in full swing, and at this point in the day, Derek and Marta are both putting functions to work for them. Functions are making their work easier. When Derek's parents drive through for a wash, Derek shows them a function that he and Marta made that tells how many volunteers they need for every hour.
>
> "See, $v(t) = t + 5$," says Derek, waving a piece of paper in his mom's face. "That means we take the number of hours we have been working and add 5 to *that* number, and we have the number of volunteers we need. Then we have $s(v) = 2v$, which means that however many volunteers we have, we multiply that by 2 to tell us how many sponges we need."
>
> Derek's mom tips him a dollar for the wash, and she and Derek's dad drive off.
>
> Marta and Mr. Ramirez walk over. "Nice tip, Derek," says Mr. Ramirez. "Let me offer you another one. You're working too hard. If you want to figure out how many sponges you need, you have to make two calculations every time. First, you use your volunteers function to figure out the number of volunteers. Then you use your sponges function to figure out your number of sponges."
>
> "It *does* take a little time," says Derek, "but so what?"

Combining and Transforming Functions

"So, you can do *half* the work if you put a composite function to work for you."

"I'm listening," Derek says.

Mr. Ramirez explains. "Your sponges function depends on the number of volunteers. Your volunteers function depends on the time of day. Put your volunteers function *in* the sponges function as though it's just another number. What do you get?"

Derek has difficulty thinking about how functions could be substituted for a variable like numbers, but Marta helps out. "It's $2t + 10$," she says.

"Right," says Mr. Ramirez. "That tells you how many sponges you need for each hour. You have made a new function—a *composite* function: $s(t) = 2t + 10$."

Extending understanding: Car-wash coupons and discounts

A second example from the car-wash context can solidify and extend students' understanding. The following vignette backtracks in time to Marta and Derek's strategizing before the car wash about how to attract new customers to this year's event. The vignette is presented in two parts, each followed by questions for students. Your students' responses to these questions will reveal their understanding and misunderstanding.

While planning the car wash, Derek and Marta approach Mr. Ramirez after algebra class to share an idea for bringing new customers to the car wash. They are planning to distribute $2-off coupons to community members who live in areas that have sent few customers to the car wash in recent years. However, the coupon strategy heightens their worry about getting too many customers at the popular times in the middle of the day. In an attempt to balance the demand across the time periods, they have decided to offer discounts for certain times of the day. Derek and Marta have organized the time periods and discounts into a chart (see fig. 3.7), which they share with Mr. Ramirez.

Discount period	Time	Discount
1	8:00 a.m.–10:00 a.m.	15%
2	10:00 a.m.–noon	5%
3	Noon–2:00 p.m.	0%
4	2:00 p.m.–4:00 p.m.	0%
5	4:00 p.m.–6:00 p.m.	10%
6	6:00 p.m.–8:00 p.m.	20%

Fig. 3.7. Marta and Derek's chart showing time periods and discounts for the car wash

Marta and Derek also ask Mr. Ramirez whether a function can represent the discounts and the coupon.

"Of course," replies Mr. Ramirez, "but this is going to require a composite function."

"What's that?" Derek and Marta say at once.

"You know most of the mathematics behind this already," says Mr. Ramirez. "Now you can build on that knowledge."

"We haven't decided on the final price for the car wash for this year," Marta says.

Mr. Ramirez merely chuckles. "That's even better," he says while the two students look puzzled.

He begins writing on the board: "Discount coupons: $2 discount off the car wash." He continues to make notes on the board while talking the problem through with Marta and Derek:

"OK, if the price of a car wash is x dollars, then $f(x) = x$, and with the $2 discount, $f(x) = x - 2$. This means that customers who do not have a coupon pay full price, in this case, x. Customers with a coupon pay $x - 2$ dollars. Now we have to consider the time of day to determine the final price."

Derek and Marta nod, and Mr. Ramirez continues:

Combining and Transforming Functions

"From your chart, I see that you've set up six time periods for the car wash, and the discount varies with those times. Hmm... I see that one discount is repeated—tell me, does that matter? Can this still be a function?"

Letting this question hang in the air, Mr. Ramirez moves on. "OK, we will use $g(x)$ for the time-of-day function. We still use x because we are seeking an expression for dollars. For the first time period $g(x) = 0.85x$, reflecting 15 percent off the price we set for a car wash. The second time period would be $g(x) = 0.95x$.

"List the function for each domain element for the discount," Mr. Ramirez suggests, "and next to those functions, list the function for the coupon."

Marta and Derek sit down at a table in Mr. Ramirez's classroom and make a chart pairing time periods and corresponding discounts. Then beside each pair they list the function for the discount and the function for the coupon (see the first four columns in the chart in fig. 3.8).

Mr. Ramirez says, "Great. Now we can put them together." He adds headings for two more columns to the right in Marta and Derek's chart, with the headings "$f(g(x))$" and "$g(f(x))$" (see columns 5 and 6 in the chart in fig. 3.8).

Time	Discount	$f(x)$	$g(x)$	$f(g(x))$	$g(f(x))$
8:00 a.m.–10:00 a.m.	15%	$f(x) = 0.85x$	$g(x) = x - 2$	$f(g(x)) = 0.85x - 1.70$	$g(f(x)) = 0.85x - 2$
10:00 a.m.–noon	5%	$f(x) = 0.95x$	$g(x) = x - 2$	$f(g(x)) = 0.95x - 1.90$	$g(f(x)) = 0.95x - 2$
Noon–2:00 p.m.	0%	$f(x) = x$	$g(x) = x - 2$	$f(g(x)) = x - 2$	$g(f(x)) = x - 2$
2:00 p.m.–4:00 p.m.	0%	$f(x) = x$	$g(x) = x - 2$	$f(g(x)) = x - 2$	$g(f(x)) = x - 2$
4:00 p.m.–6:00 p.m.	10%	$f(x) = 0.90x$	$g(x) = x - 2$	$f(g(x)) = 0.90x - 1.80$	$g(f(x)) = 0.90x - 2$
6:00 p.m.–8:00 p.m.	20%	$f(x) = 0.80x$	$g(x) = x - 2$	$f(g(x)) = 0.80x - 1.60$	$g(f(x)) = 0.80x - 2$

x = number of customers

Fig. 3.8. For different times and discounts, corresponding price-after-discount functions ($f(x)$), price-with-coupon functions ($g(x)$), and composite final-price functions ($f(g(x))$ and $g(f(x))$) for the car wash

Mr. Ramirez continues talking, guiding Marta and Derek as they fill in the new columns.

"In the first time period, 8:00–10:00 a.m., $f(x) = 0.85x$ and $g(x) = x - 2$. The composite function is $f(g(x))$, which means that to evaluate the f function, you must first find $g(x)$. In this case, $g(x)$ serves as a variable of the function f. So, $f(g(x)) = f(x - 2)$ since $g(x) = x - 2$. And since $f(x) = 0.85x$, $f(x - 2) = 0.85(x - 2)$, which simplifies to $0.85x - 1.70$."

At this point, posing some questions may help you assess students' understanding. (Now might also be a good time to pursue with students the question that Mr. Ramirez poses to Marta and Derek about whether the fact that one discount [0 percent] repeats over two time periods affects the status of the relationship as a function.) However, before posing the questions, note that using $g(x)$ as a variable in $f(x)$ requires that students' understanding of the idea of a variable be at level 6 of Kuchemann's [1978] hierarchy.

1. What property is used to simplify $0.85(x - 2)$?

2. In the function, $f(g(x)) = 0.85x - 1.70$, what does 1.70 mean in the context? What does the 0.85 mean? How does this apply to other functions in the list?

3. How would the composite function represent a customer without a coupon?

4 How would the function change if coupons were worth $2.50?

Before returning to the concluding section of the vignette, pause in your reading to consider the questions in Reflect 3.5.

Reflect 3.5

How would you answer the questions posed above for students about the composite final-price functions (in fig. 3.8) for the car wash?

How would you expect your students to answer them?

Like Marta and Derek, your students may feel that their understanding about composing functions is growing. But some confusion may persist, as in this case:

Marta and Derek are excited, feeling confident that they are close to understanding how to handle the discount periods and the coupon for the car wash.

"Can we now substitute the cost of the car wash for x when we finalize the price?" asks Derek.

"Wait," Marta says. "First, how do we know which composite function to use, $f(g(x))$ or $g(f(x))$?"

"Perhaps you should try both and see what you learn," replies Mr. Ramirez.

"OK," Marta says to Derek. "How much do we want to say that x, the cost of a car wash, is going to be?"

"Well, earlier we were thinking $10, but now I'd say maybe $12," says Derek. "What do you think?"

Marta agrees, and they quickly begin to work, creating a chart to show their results (see fig. 3.9). When they are finished, they study their new chart for quite a while.

Time of day	Discount	$f(g(x)))$	$g(f(x))$
8:00 a.m.–10:00 a.m.	0.85	$ 8.50	$ 8.20
10:00 a.m.–noon	0.95	$ 9.50	$ 9.40
Noon–2:00 p.m.	1	$ 10.00	$ 10.00
2:00 p.m.–4:00 p.m.	1	$ 10.00	$ 10.00
4:00 p.m.–6:00 p.m.	0.9	$ 9.00	$ 8.80
6:00 p.m.–8:00 p.m.	0.8	$ 8.00	$ 7.60

Fig. 3.9. Function values for discount and carwash costs for $f \circ g$ and $g \circ f$ with $x = \$12$

Ask your students the following questions to focus their attention both locally and globally on the composed functions:

1. What can you conclude from Marta's and Derek's work (shown in fig. 3.9)? Describe the differences in output for $(f \circ g)(x)$ and $(g \circ f)(x)$.

2. Will $(g \circ f)(x)$ always be less than or equal to $(f \circ g)(x)$? Can you think of a counterexample?

3. Which composition should Marta and Derek use, and why?

4. How will the values in the table change if Marta and Derek change the price of the car wash to $15? To $8? Will $(g \circ f)(x)$ remain less than or equal to $(f \circ g)(x)$ in every case?

5. Describe how these results would change if the "discount" actually became a "premium cost" of 1.25 percent for the hours between 12:00 noon and 4:00 p.m. Explain these different results.

Further explorations in combining and composing functions: Prom Room Rental, Delivery Fees

Two school-based problems offer students opportunities to explore functions and their combination and composition further. The first problem involves renting a room for the junior/senior prom, and the second revisits the car-wash setting to examine delivery charges for advertising materials. Each problem asks for a function to represent the situation. Students should discuss how their function captures the parameters of the particular story. Urge them to explain how they determined values for domains and ranges that would be reasonable for that particular application.

Prom Room Rental

The junior class organizes the junior/senior prom each year. Class members discover that the community banquet hall charges $500 to rent, plus $15.45 for each person attending an event. If they book the hall this month, they will receive a discount of 30% off the total bill. Write a function to model each of these cost situations. Combine these two functions to show the cost as a function of the number of students attending the prom.

(Note that students might write the symbolic expression $f(p) = 15.45p + 500$ to represent the before-discount cost as a function of the number of students attending. Combining this function with the 30% discount would give them $f(p) = 15.45p + 500 - 0.3(15.45p + 500)$, or $f(p) = 0.7(15.45p + 500)$.)

Delivery Fees

Jamie and Denitha make a purchase at Harding Hardware to build a standing display to highlight the car-wash site. The collection of materials that they purchase is too big to take home in their car. The hardware store will ship the materials to the site for a special school rate of $15. Jamie and Denitha pay for the purchase, along with the 6.5% sales tax and the fee.

1. Write a function $t(x)$ for the total cost, including tax but excluding the delivery fee, for the purchase amount x.

2. Write another function $f(x)$ for the total cost, including the delivery fee but excluding the tax, for the purchase amount x.

3. Calculate and interpret $(f \circ t)(x)$ and $(t \circ f)(x)$. Describe the results and what they mean.

4. If the sales tax increased to 7%, how would your results change? Describe the change if the normal delivery fee of $50 were charged.

As students determine the domain and range under function composition, their misconceptions may become evident. The composition of two functions exists only if the range of the first function is contained in the domain of the second function. The domain of $f \circ g$ is a subset of the domain of g. The range of $f \circ g$ is a subset of the range of f. Figure 3.10 diagrams the composition of f and g. Note that $g(x)$ must be defined so that any x not in the domain of g must be excluded, and $f(g(x))$ must be defined so that any x for which $g(x)$ is not in the domain of f must be excluded.

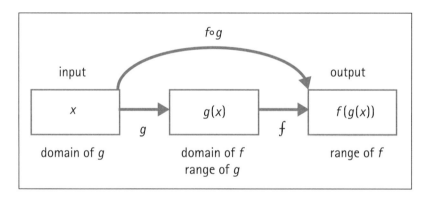

Fig. 3.10. Diagram of composition of functions f and g

Transformations

Combining or composing functions as new functions changes their graphs in predictable ways. Transforming functions by reflections, translations, rotations, and dilations also produces predictable results. Students may enter high school understanding these transformations as flips, slides, turns, and stretching and shrinking, but teachers should help their students build on these terms as they move to more formal language, as found in Common Core State Standards (National Governors Association Center for Best Practices and Council of Chief State School Officers 2010). The examples that follow illustrate ways to work with students on ideas related to transformations of functions.

Issues of scale: Painting the mascot

At the center of the following vignette is a transformation that results from a change of scale. Again, the vignette features Mr. Ramirez and his students, but this time they are in Mr. Ramirez's algebra classroom, and the project under discussion is a new one: the creation and painting of an enlarged image of the school's mascot—a dilation of a small, familiar iconic image in the school:

> "There is another way of representing a function," says Mr. Ramirez. "We have seen functions represented as tables. We have seen them represented as graphs. Now you are seeing a function represented with symbols. Each of those representations is useful for different things. We will talk more about symbols later, but first I want to see how well you understand function families and their graphs."
>
> Derek says, "By function families, you mean linear and quadratic, exponential and periodic functions."
>
> Mr. Ramirez replies, "Yes, but not all at once. We have another class project that might help us with this idea. Who is part of the group planning to paint the new tiger mascot mural in the school's main hallway?"
>
> The five spirit club members in the class raise their hands.
>
> Students at Grover Cleveland High School are proud of the painting of their mascot over the entry door. This small image (see fig. 3.11) is 56 centimeters high and located just below the school's name. The club has proposed that a large mural (see fig. 3.12) 2.8 meters high be painted in the main hallway just outside the school cafeteria. The art club has offered to create the mural.
>
> "Well," says Mr. Ramirez, "how much paint will be needed in all? Of each color?"
>
> Fortunately, the school has a record of the amount of paint used for the original painting (see the table in fig. 3.13; the left-hand column indicates original colors and color substitutions in the figure).

Combining and Transforming Functions

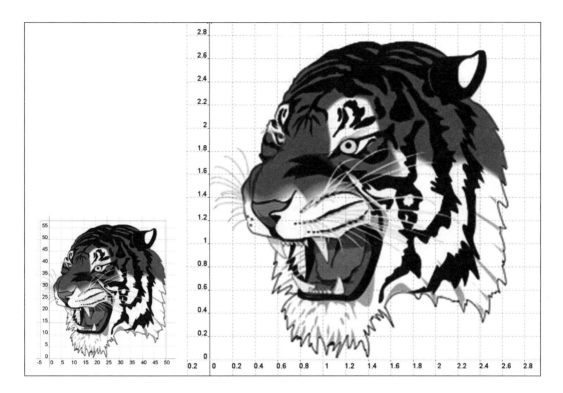

Figure 3.11. Small tiger image (cm)   Figure 3.12. Proposed large tiger mural (m)

Paint used in original image	
Color	Amount (ml)
White (as shown in fig. 3.11)	4.7
Orange (richest shade in the figure)	2.3
Yellow (for tiger's eyes; light shade in the figure)	2.5
Black (as shown in the figure)	2.5
Red (for tiger's tongue, gums, nose; medium shade in the figure)	0.7
Ivory (for tiger's teeth; palest shade in the figure)	0.3
Total	13.0

Fig. 3.13. A table showing the milliliters of paint used for the small image of the school mascot

The task for your students is the same as for the students in the scenario:

> Given the amounts of different colors of paint that the school used in creating the small image of the mascot, how much paint will the students need in total for the new, large mural showing the same image? How much white paint will they need?

In a research study, De Bock and colleagues (2002) offered students a similar problem. They discovered that students most often applied linear reasoning to solve the problem. That is, the students reasoned that if the figure were five times bigger (taller), then the new image would require five times the amount of paint. De Bock's team followed up with leading questions and additional scenarios to determine the robustness of this misconception. The following questions and strategies slightly modify those posed by De Bock and colleagues, adjusting them to fit the tiger-mural scenario. De Bock and associates ordered their questions to address phases of error in students' thinking, with successive questions targeting the error persistently and repeatedly, as necessary. The questions and strategies that follow correspond to their phases 2–5. Direct the questions to your students after they have read the scenario:

> Phase 2: Last week, students in another school worked on the same problem and got mixed results (see the table in fig. 3.14). How might one group of students at that school have arrived at 325 milliliters as the total amount of paint needed?
>
> Phase 3: One student said that if the height is multiplied by 5, the width must be multiplied by 5, so the amount of paint must be multiplied by 25. Do you agree? Why or why not?
>
> Phase 4: Draw a rectangle around each of the two tiger figures.
>
> Phase 5: (Structured consecutive questions)
>
> You've drawn rectangles around both tiger images. How would you determine the area of the two rectangles?
>
> How much larger is the large rectangle than the small one?
>
> How much larger is the area of the large tiger than the small one?
>
> How much more paint will you need to paint the large tiger than the small one?

Responses from students at another school	
Total paint (ml)	Percentage of students
65 ml	43%
325 ml	43%
Other	14%

Fig. 3.14. Responses of fictitious students at another school to the problem about the total amount of paint needed for the enlarged image

At phase 2, De Bock's team attempted to provoke a mild form of cognitive conflict in students who solved the problem incorrectly. Given some time to reflect on the other class's responses, students were asked which answer they preferred: their initial answer or the alternative from the other class. Students who did not abandon their initial response then moved on to the phase 3 question. The argument in the question at this phase came from a fictitious student who was in the 43 percent of students who answered the problem correctly (325 ml). Students were again asked if they wanted to stick with their original answer (linear) or the alternative (nonlinear). Students who did not change their original answer moved on to the phase 4 prompt. By asking the students to draw rectangles around the figures, this prompt was designed to trigger greater cognitive conflict by demonstrating the reasoning behind the correct answer (325 ml). The set of consecutive questions at phase 5 prodded students who were still in error to focus explicitly on the areas of the figures.

The tiger-mural activity and the accompanying set of structured questions are designed to investigate the degree to which students are committed to the misconception that functions are linear or that relationships in an application need to be linear. By completing the following tasks, students continue to reveal their understanding or misunderstanding:

1. Given the record of the amount of paint, c, required for the small image of the mascot, create a function that will yield the amount of paint needed for the large image.

   [The functions that the students write will show their thinking:

   $f(c) = \dfrac{2.8}{0.56} c$, or $f(c) = 5c$ (the common misconception)

   $f(c) = 5^2 c$, or $25c$ (the accurate representation)]

2. How would the function change if the size of the larger mural were 3.08 meters? 2.24 meters?

3. How are the graphs of these three functions (for the amounts of paint for images of 2.8, 3.08, and 2.24 meters) the same? How are they different?

Transforming with coefficients or constants

When changing a coefficient or a constant in a symbolic representation of a function, students sometimes struggle to predict the impact that the change will have on the graph of the function (Borba and Confrey 1996). Students can gain experience with these concepts by using interactive graphical representations to explore multiple examples of linear and quadratic functions expressed in the following general notation:

$$\text{Linear functions: } f(x) = mx + b$$

$$\text{Quadratic functions: } f(x) = a_2 x^2 + a_1 x + a_0$$

Students might approach questions about the impact of changes in coefficients or constants by developing tables of values or by graphing with GeoGebra (http://www.geogebra.org/cms/en), as illustrated in figure 3.15, or with another tool capable of graphing functions.

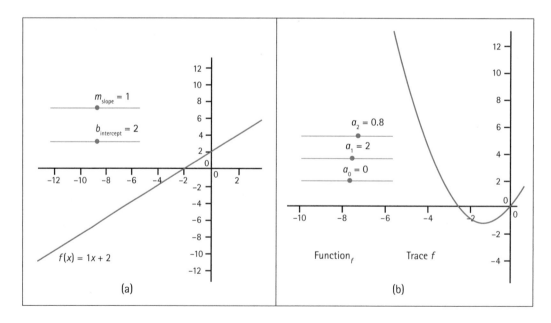

Fig. 3.15. GeoGebra sketches with sliders for (a) the linear function $f(x) = 1x + 2$ and (b) the quadratic function $0.8x^2 + 2x + 0$

Combining and Transforming Functions

Key questions to pose to students about the functions and their graphs include the following:

1. How does changing the values of m and b in linear functions affect the shape and location of the graph of the function?

2. Describe how the values a_2, a_1, and a_0 change the shape and location of a quadratic equation.

3. Describe some values of m and b that indicate special properties of a linear function.

4. Describe values of the coefficients of quadratic functions that indicate special situations—for example, $a_1 = 0$.

5. Compare and contrast the impact that the coefficients of the linear and quadratic equations have on the graph.

Applications for translations: Car-Wash Prices, Water's Height, Water's Surface Area

The following three tasks offer opportunities to reinforce students' understanding of transformations and extend their thinking about translations. Commentary accompanying the second and third tasks provides guidance for using them with students; the first task appears on its own, without discussion.

1. **Car-Wash Prices**

    Recall the pricing structure that Marta and Derek work out for their club's upcoming car wash. They plan to distribute $2-off coupons to all customers and offer additional discounts that vary according to the time of day. With x as the base price of a car wash (in dollars), they write functions $f(x)$ for the price after the discount and $g(x)$ for the price with the coupon. With Mr. Ramirez's help, they compose their price-after-discount and price-with-coupon functions to write functions for the final price of a car wash (fig. 3.16 reproduces fig. 3.8 for the reader's convenience).

Time	Discount	f(x)	g(x)	f(g(x))	g(f(x))
8:00 a.m.–10:00 a.m.	15%	f(x) = 0.85x	g(x) = x − 2	f(g(x)) = 0.85x − 1.70	g(f(x)) = 0.85x − 2
10:00 a.m.–noon	5%	f(x) = 0.95x	g(x) = x − 2	f(g(x)) = 0.95x − 1.90	g(f(x)) = 0.95x − 2
Noon–2:00 p.m.	0%	f(x) = x	g(x) = x − 2	f(g(x)) = x − 2	g(f(x)) = x − 2
2:00 p.m.–4:00 p.m.	0%	f(x) = x	g(x) = x − 2	f(g(x)) = x − 2	g(f(x)) = x − 2
4:00 p.m.–6:00 p.m.	10%	f(x) = 0.90x	g(x) = x − 2	f(g(x)) = 0.90x − 1.80	g(f(x)) = 0.90x − 2
6:00 p.m.–8:00 p.m.	20%	f(x) = 0.80x	g(x) = x − 2	f(g(x)) = 0.80x − 1.60	g(f(x)) = 0.80x − 2

x = number of customers

Fig. 3.16. Price-after-discount functions (f(x)), price-with-coupon functions (g(x)), and composite final-price functions (f(g(x)) and g(f(x))) for different car wash times and discounts. (Reproduces fig. 3.8 for the reader's convenience.)

    a. Graph the lines of the composite functions and compare the differences.

    b. How do the intercepts of the lines relate to the information in the table?

    c. How do the slopes of the lines relate to the table?

    d. How does the discount change the functions? Are there differences between $f \circ g$ and $g \circ f$?

    e. How does the initial cost of the car wash change the functions? Does it have a greater impact on $f \circ g$ or on $g \circ f$?

    f. If only positive values are entered into one of these composite functions, is it possible to obtain a negative value with the function?

2. **Water's Height in a Trough**

A water trough (see fig. 3.17) is 10 inches wide at the base and 10 inches deep. It has a right angle at the back and a 135-degree angle at the front. Water is filling the trough at a constant rate. When the water reaches a height of 4 inches, the width of its surface on the end of the trough is 14 inches.

Combining and Transforming Functions

What is the function that shows the rate of change of the water's width, given its height?

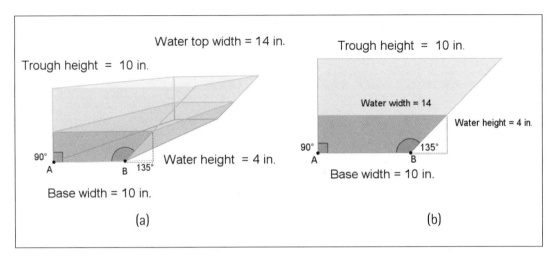

Fig. 3.17. Two perspectives on the trough with the water's height at 4 inches: (a) a projective sketch and (b) an end view

The water trough problem brings together in one problem a number of ideas about combining, composing, transforming, and interpreting functions. First, students must develop a function that captures the rate of change of the water's width, given its height. Follow-up questions require changing the developed function in concert with changing conditions (shape) of the trough. (Note that you can also simplify the problem for your students by making both the front and the back angles of the trough 90 degrees.)

Students begin by developing a function that will determine how the width of the water in the trough changes as the height changes. The front face of the trough forms a 135-degree angle with the base, which means that in the end view shown in figure 3.17b, the external triangle formed by dotted lines is a right isosceles triangle. Hence, the width of the water surface changes at the same rate as the height (since the base of the triangle is equal to its height). Therefore, the width, $f(h)$, of the water, given h, its height in the trough, can be expressed as $f(h) = h + w_0$, where w_0 is the initial width of the water in the trough, or the width of the base of the trough. So $f(h) = h + 10$ gives the width of the water as a function of the height h.

Students can use GeoGebra to create a series of diagrams to explore how the function changes if the width of the base of the trough changes. Figure 3.18 shows a set of four such explorations, with the base varying from 5 to 20 inches for functions f_1 to f_4.

Putting Essential Understanding of Functions into Practice in Grades 9–12

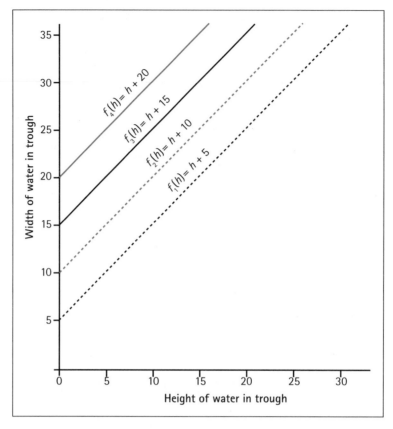

Fig. 3.18. Graphs of the water's width in the trough as a function of its height, h, for changing widths of the base of the trough

After creating the graphs in the figure, your students can respond to a number of questions to help them analyze the representations more fully and compare them more meaningfully:

1. What do the four functions describe together? How does changing the width of the trough change the function?

2. Why are the lines for the four functions parallel? Describe the parallelism in terms of the rising water.

3. What happens at $h = 0$?

4. What is changing in the functions? What stays the same? Describe what changes and what stays the same in terms of the water in the trough and in terms of the graph.

5. How would the function change if the base were zero?

6. How would the function change if both sides of the trough were set at 135 degrees (see fig. 3.19)? At 90 degrees?

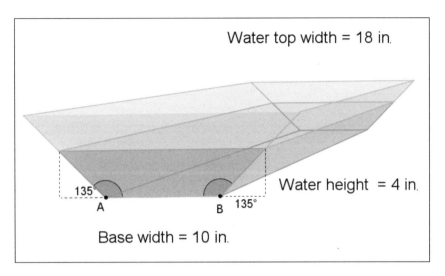

Fig. 3.19. Water trough with front and back sides forming 135-degree angles with the base

7. What are the limits of the domain and range for these functions?

In responding to question 6 about the function in the case of the trough pictured in figure 3.19, students should reason that since both faces of the trough form 135-degree angles with its base, two external isosceles triangles are formed instead of just one, as before, and these triangles are congruent, so the increase in the width of the water in the trough is now twice the increase in the height.

Understanding the value of $f(h)$, the water's width at height h, in this set of functions may be difficult for some students at $h = 0$. These functions appear to have a y-intercept with a value greater than zero; that is, at $h = 0$, $f(h) = 10$ (for a trough with a width of 10 inches). Yet, if there is no water in the trough, there is no height or width. However, when there is enough water in the trough to measure height, it will cover the bottom and then the width will be zero.

Using the same model, students can gain experience with more advanced functions by exploring the relationship between the surface area of the water, either on an end or a face of the trough, and its height. Exploring the surface area of the water on the trough's end, presented next as task 3, involves quadratic rather than linear functions.

3. Water's Surface Area on the End of a Trough

A water trough (see fig. 3.20) has a base of 10 inches and a height of 10 inches. Both the front and the back side of the trough form a 135-degree angle at the base of the trough. Water is filling the trough at a constant rate and has reached a height of 4 inches and a width of 18 inches across the end of the trough.

How does the surface area of the water on the end of the trough change with the height?

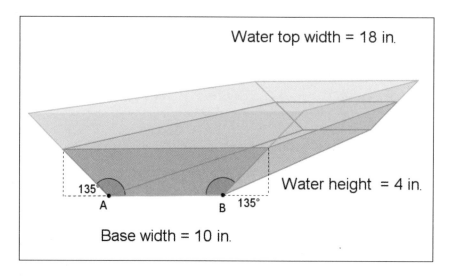

Fig. 3.20. A trough with a base and height of 10 inches, with front and back sides forming 135° with the base

The surface area of the water on the face of the end of the trough can be modeled as a function of the height, h, of the water in the trough: $a(h) = h^2 + 10h$ (or, in general, $a(h) = h^2 + hb$, where b = width of base).

Students can then explore how the representation of this function changes as the base of the trough changes by comparing the function notation and the graph with the story that they both model. Figure 3.21 shows three functions that model the surface area covered by water on the end of the trough as the height changes for troughs of different base sizes.

Combining and Transforming Functions

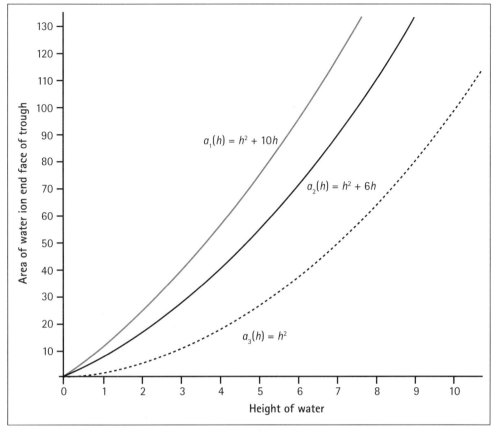

Fig. 3.21. Graphs of surface area of the water on the end of a trough as a function of the water's height, h, for changing sizes of trough bases

The following questions may help guide students to analyze these representations more fully and compare them with more meaning:

1. What do these functions describe together?

2. How does changing the width of the base of the trough change the function?

3. What happens at $h = 0$?

4. What is the value for each of the functions at $h = 1$? What does this mean in in terms of water in the trough?

5. What is changing in the functions? What stays the same? Describe what is changing and what stays the same in terms of the water in the trough and in terms of the graph.

Putting Essential Understanding of Functions into Practice in Grades 9–12

6. How would the function change if both sides of the trough were changed?

7. What are the limits of the domain and range for these functions?

Students can explore additional shapes for the trough to investigate the impact on the symbolic and graphical representations of the respective functions. Attending to the domains and ranges of these functions is important and may help students connect these representations more closely with the respective stories. A few possible shapes for consideration are shown in figure 3.22.

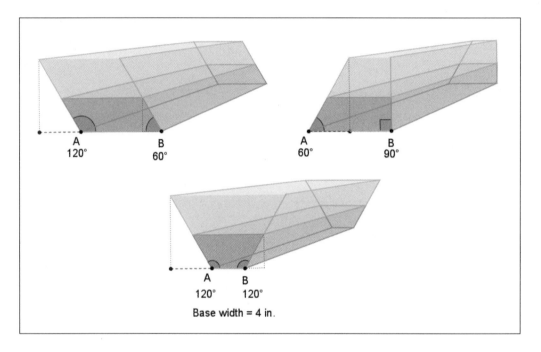

Fig. 3.22. Three additional water trough shapes for exploration

After developing functions for the surface area of the water on the end of the trough, students can enhance their understanding by focusing on the area of the top surface of the water in the trough, an extension that would bring the length of the trough into play. Students can also explore how the volume of the water changes with the change in its height.

Conclusion

In thinking about the examples presented in this chapter, note that at this point in their development, your students are operating with functions as though the functions were themselves objects, rather than a list of operations or a prescription for turning an input value into an output value. They are adding *functions*, whereas before they added only real numbers. They are multiplying with notation like "$f(x)$" and "$g(x)$," whereas before they multiplied only by using numbers and times tables. These are not trivial shifts. At certain points in their processes of combining, composing, and transforming functions, students may lapse into procedural operations—adding like terms, reducing rational expressions to lowest terms, and multiplying monomials. At these moments, you may need to draw their attention from the abstract realm of the symbolic algebra to the concrete situation that gave rise to the abstraction. Ask questions about each of the coefficients in the combined function: "What do these numbers mean in terms of the story? Where do you find them on the graph?" And, of utmost importance: "Which numbers no longer work? What is the new domain of the combined function?" Domain and range remain important concepts that you should stress throughout these explorations.

Sample questions have accompanied each of these scenarios, but the task is yours to alternate the focus between local values and global features like the domain and range, and between the abstraction of function notation and the concrete instantiation of that notation in the world itself.

into practice

Chapter 4
Graphs as Representations of Functions

Essential Understanding 5a
Functions can be represented in various ways, including through algebraic means, graphs, word descriptions, and tables.

Essential Understanding 5d
Links between algebraic and graphical representations of functions are especially important in studying relationships and change.

Big Idea 5 in *Developing Essential Understanding of Functions for Teaching Mathematics in Grades 9–12* (Cooney, Beckmann, and Lloyd 2010) has two parts. The first is the notion that "functions can be represented in multiple ways, including algebraic (symbolic), graphical, verbal, and tabular representations" (p. 8), and the second is the entwined insight that "links among these different representations are important to studying relationships and change" (p. 8). This chapter focuses on the first part of this big idea, which Cooney, Beckmann, and Lloyd identify as Essential Understanding 5a, and it also gives some special attention to the second part, which those authors identify as Essential Understanding 5d. To begin, consider the following vignette, which returns once again to the car-wash setting:

> It is eight o'clock. The car wash is over. Derek and Marta planned well. Two bottles of soap and four sponges are all that remain of their supplies. Marta is standing next to the remaining supplies, scowling at a piece of paper. On it is a graph that Derek has made.
>
> Mr. Ramirez notices her expression. "What's the problem, Marta?" he asks.
>
> Marta says, "Derek left me this graph of the cars that we washed by ourselves from the end of the first hour after we opened until six o'clock, when

we rotated off the washing team. He's bragging that he washed more cars than I did, and he says this graph proves it." (Fig. 4.1 shows the graph.)

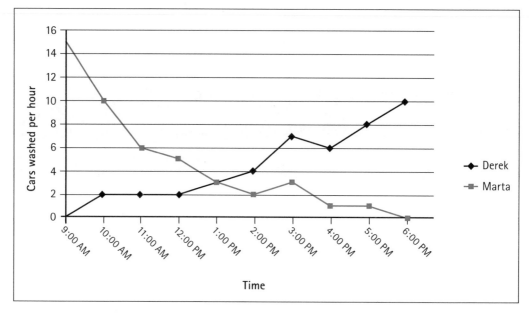

Fig. 4.1. Graph of the cars washed by Derek and Marta

"He told me that the number of cars I washed actually went down over time," Marta continues. "But how can that be possible?"

"I think I know what you are seeing," says Mr. Ramirez. "Let me try to explain."

You may have found yourself in analogous situations in the classroom, facing students who are similarly puzzled. Reflect 4.1 provides an opportunity for you to consider how you might respond to Marta.

Reflect 4.1

What would you say to Marta if you were Mr. Ramirez?
How would you help her understand the situation?

Graphs as Representations of Functions

This chapter explores a set of graphs that represent functions that model various relationships. The selected functions highlight specific student misconceptions that research has reported. The examples are accompanied by possible questions for students and suggestions about how they might respond, depending on the particular misconceptions that they have.

The use of multiple representations has been recommended as a means of deepening students' understanding of concepts and the relationships among those concepts (National Council of Teachers of Mathematics 2000; Kaput 1989). This chapter emphasizes understandings and misconceptions of functions represented as graphs and ways of connecting these representations with other types of representations. Although Essential Understandings 5a and 5d are at the center of this chapter, some misconceptions do not reflect a failure to grasp a single essential understanding or big idea. The discussion that follows includes some instances of these.

Different representations of functions provide distinct insights into the relationships that they model. Analyzing and applying these representations are critical learning strategies for gaining insight into and making sense of the models. Although the various representations of a particular function may look very different, students must understand that the covariance relationship that they represent remains constant (Cooney, Beckmann, and Lloyd 2010).

Addressing Misconceptions about Graphs

A key representation for functions is graphs. To make use of graphs of functions, students must be able to understand them and recognize how they represent the covariance of the function. The following examples illustrate some misconceptions that students frequently demonstrate while attempting to interpret graphs.

Velocity of One Car

A number of researchers have investigated misconceptions that students exhibit in working with functions and their graphs (Leinhardt, Zaslavsky, and Stein 1990; Monk 1992, 1994, 2003; Clement 1989; McDermott, Rosenquist, and van Zee 1987). Leinhardt and colleagues (1990) point out that algebraic representations of functions and graphical representations are two very different symbol systems that students must coordinate conceptually to make sense of a relationship represented in these ways. They cannot treat these representations as isolated concepts. Moreover, working from graphical representations and starting from algebraic representations are two very different sense-making processes. Further, students' intuitions about contexts or situations represented in a graph may contribute to their misconceptions about the representation (Monk 1992; Leinhardt, Zaslavsky, and Stein 1990).

A graph of a function provided by Monk (2003, p. 251) offers a simple example that shows some of the complexity of this concept. Monk's graph, presented here as figure 4.2, shows the velocity of a single car over a period of 5 minutes.

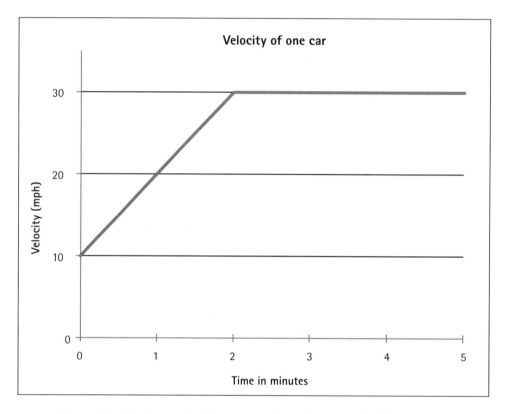

Figure 4.2. Velocity graph of one car. Adapted from Monk (2003), p. 251.

Monk identifies a number of ways in which this graph may be misinterpreted—misreadings that are associated with viewing it as a picture of an event rather than as a type of analysis of the event. For example, some learners might think that the car stops at $t = 2$, even if they realize that the vertical axis represents velocity, not distance or position.

A key skill for interpreting such graphs is to be able to screen out attributes that are not needed to understand the graph. Leinhardt and colleagues (1990) identify two critical foci—local and global—for interpreting graphs, as we have noted and discussed previously. On the one hand, a global focus is more qualitative by nature and addresses the overall relationship between the variables of the function. This focus is on the meaning of the graph, as conveyed by the "big picture." On the other hand, a local focus fixes on individual components of the graph—its slope, height, range, and so on. Understanding a graph often requires both perspectives. However, instruction in earlier grades does not typically address a global treatment of graphs, which thus becomes an important skill for secondary teachers to help students develop.

Stated concisely, interpretations of graphs are of two general types—qualitative and quantitative—corresponding to global and local perspectives. A qualitative interpretation focuses on the overall picture of the graph to comprehend the relationship or covariance of the two variables (a global perspective), whereas a quantitative interpretation seeks to extract quantities from the graph (a local perspective) (Leinhardt, Zaslavsky, and Stein 1990).

For Monk's graph of the velocity of one car, a qualitative question might focus on the shape of the graph, perhaps seeking a comparison of meaning of the slanted and the horizontal line segments or an explanation of the meaning of the "straightness" of the segments. A quantitative question might require a comparison of the velocity of the car at times 1, 3, and 5 or an explanation of what happens at $t = 2$. (Some learners might suggest that the car stops at $t = 2$; even learners who realize that the y-axis represents velocity rather than distance or position sometimes give such a response.)

The following questions about the graph in figure 4.2 require both an internal (local, quantitative) focus on specific components of the graph and an external (global, qualitative) focus on the coordinate system and the overall meaning of the graph:

1. What do the two line segments in the graph indicate? Compare the velocity represented by the oblique (slanted) segment and the velocity represented by the horizontal segment. (global)

2. People may misinterpret the meaning of the straight line segments in the graph. Do these straight segments indicate no change? How does the oblique segment differ from the horizontal segment in meaning? What would the car have to do for these lines in the graph to be more "wavy?" (global)

3. How would you compare the car's action at $t = 0$ and its action at $t = 3$? What does it mean that the graph starts at a velocity of 10 miles per hour?

Can a car be going 10 miles per hour at time $t = 0$? What can you say about the velocity of the car at $t = 6$? (local)

4. What happens at $t = 2$? Does the car stop? What is the change at $t = 0$? Does the car stop? Write a short story (just a few lines) describing what it would be like if you were in the car represented by this graph. (local and global)

Because of the two distinct perspectives—local and global—that interpretation of this graph invites, it can help students learn to use graphs as dynamic mathematical representations that assist in making sense of a covarying relationship. This particular illustration is useful for building or diagnosing students' understanding of graphs because of its simplicity—that is, the graph consists of one line (a broken line) with simple values.

At the same time, the graph contains sufficient complexity to require some thought to interpret it at both the global and local levels. Examples representing this and similar simple scenarios may also be acted out; in this case, a reasonable pace might be something like 3 miles per hour, or 1 step per second.

Once students are able to make meaningful local and global interpretations of graphs like this one, they should be able to construct additional representations, such as tables or narratives. You can further assess your students' understandings of these key ideas by constructing, or having them construct, graphs from examples of non-graphical function representations and then having them interpret the graphs. Facility with these examples may indicate a readiness to interpret graphs representing two distinct functions.

Comparing Two Cars, Given Distance

Monk (2003) designed the preceding example of the velocity of a car to assess and provide instruction to address the misconception that a graph may be interpreted as a picture rather than as a dynamic tool that represents the covariance between two variables. A different example offers an opportunity to determine whether students with some notion of the meaning of graphs and a beginning understanding of covariance can use that knowledge to interpret a more complex relationship.

McDermott, Rosenquist, and van Zee (1987) studied student difficulties in connecting graphs with physical concepts. Their findings indicated that the difficulty experienced by students is related less to the physical experience presented in the scenario than to the students' inability to interpret graphs. In inspecting a straight-line graph such as that shown in figure 4.3 of the speeds of two cars, students may extract information by focusing on the coordinates of the points, the differences between the coordinates of the points, or the slopes of the line. Students may struggle to determine which feature of the graph they should use to extract information.

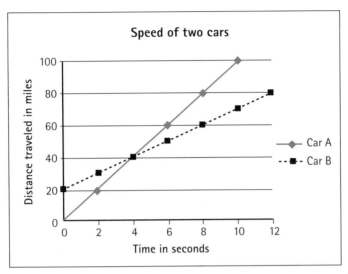

Fig. 4.3. A representation of the distances from a given location of two cars traveling in the same direction on the same road. Adapted from McDermott, Rosenquist, and van Zee (1987).

This example differs from the earlier graph of the velocity of one car in that the new graph shows a comparison between two cars rather than a representation of just one car. Having a second car creates a new level of difficulty. Further, in the new situation, the vertical axis represents distance rather than velocity, and the graphs for both cars in the new illustration are straight lines instead of the broken line that simplifies interpretation in the earlier case of one car. Questions probing students' understanding of such matters as the distance traveled by car A or car B at a time t are a logical beginning.

Moreover, some of the same questions suggested in the case of the first example can be asked in the case of this example as well—for instance, what is happening to either car A or car B at, before, or after a particular time, and what is the velocity of the car at time t? McDermott, Rosenquist, and van Zee (1987) show that students often confuse distance with velocity when comparing two objects moving in this manner.

After offering students initial warm-up questions, posing a number of comparison questions can help them explore and dig deeper into the graph. Consider the following questions, which are interspersed with observations about students' possible responses:

1. Describe what is happening at $t = 2$. Which car is going faster? How can you tell? Which car has gone farther? How far apart are the two cars?

   Learners may mistakenly think that car B is going faster when $t < 2$ because the line that represents car B is above that for car A. Some learners may indicate that at $t = 2$ car B has gone farther, either because the graph for car B is above the graph for car A or because they are confused about the situation at $t = 0$.

2. Are the cars ever moving at the same speed? At what time? Explain how the graph represents this event.

   Learners have reported that at $t = 4$ the cars are going the same speed. These students fail to recognize that since the slopes of the graphs for the two cars are never equal, their speeds can never be the same. In this case, learners may not recognize that in the graph "height" represents distance.

3. Describe the location of the cars at $t = 0$. What can be said about the speeds of the cars at this time?

   Car B does not start at the same place as car A. This is sometimes confusing for learners. Which car is traveling faster at $t = 0$? Compare the locations of the two cars at $t = 0$. Important concepts for learners to grasp are that all graphs do not intersect the origin, and that they must interpret what it means if a graph does not (or does). In this case, the meaning is that car B has a 20-mile head start.

4. Does one car ever pass the other? At what time?

   This question is worth thinking about before posing it to students. Reflect 4.2 provides an opportunity for you to consider this question along with another one, presented as question 5 below.

Reflect 4.2

Suppose that you are using the example of the two-car graph in your classroom and a student asks, "Does one car ever pass the other? Do the two cars ever travel the same distance? At what time?"

How would you respond to this student?

At $t = 4$, both cars are 40 miles from the given location, so indeed they will pass each other. Look for learners who answer this correctly but incorrectly indicate that the cars are traveling at the same *speed* at $t = 4$. A correct local interpretation of the positions of the cars at $t = 4$ may not ensure that the learners correctly interpret the cars' speeds, which remain constant—and different. Interpreting the graph to give a correct response to this question depends on a global understanding.

5. Do the two cars ever travel the same distance? At what time?

   Learners who suggest that at $t = 4$ the cars have traveled the same distance are focusing on the coordinates of t (local interpretation). However, at $t = 4$, car A has traveled 40 miles, and car B has traveled 20 miles. Learners recognize this fact in a global interpretation. In the graph, the only time at which the cars have traveled the same distance is at $t = 0$. A question that might encourage the integration of local and global views is, "At what times have the two cars traveled the same distance?" This question could elicit the response that at $t = 4$ car A has traveled 40 miles, and at $t = 8$ car B has traveled 40 miles.

Distinguishing between the questions that require a focus on local characteristics and those that require a focus on global characteristics may help students who appear to struggle. You can use additional examples to assess your students' understanding and provide them with experience in interpretation. For instance, in the graph in figure 4.3, you might change the vertical axis from "Distance traveled in miles" to "Speed in miles per hour," as in the next example. Or you might modify the values to make the example appropriate for students to model by stepping off the distances on a graph on the floor.

This example requires students not only to interpret speed from a graph that displays time and distance, but also to compare the unit rates (speeds) of the two cars. Students should recognize that the unit rates remain constant for both cars throughout the graphical representation. Working with the example of the two cars should strengthen students' skill in making various local and global interpretations of graphs like this, and they should be able to construct additional representations and move fluently among them.

Comparing Two Cars, Given Speed

The next example invites examination of the speed of two cars through a different lens. Suppose that two cars, A and B, start at the same time and place and travel on the same road for one hour. The trips for car A and car B are shown in the graph in figure 4.7. Clement (1989) and Monk (1994) report that students who can provide accurate readings of the speeds of the cars at $\frac{1}{4}$-hour intervals (local perspective) often struggle to describe the relative positions of the cars at a given time (global perspective).

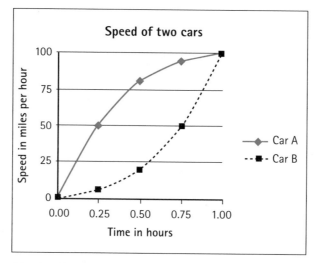

Fig. 4.4. Comparing two cars by a different lens: speed

In this example, students do not typically find the time scale on the x-axis to be nearly as troublesome as speed on the vertical axis. Some students may interpret the vertical axis as distance in some contexts and as speed in others. As before, we provide some questions interspersed with discussion of likely responses from students:

1. Which car is traveling faster at $t = 0.75$ hour?

    Learners can typically interpret this value and correctly read the speed as the height of the graph. However, this correct response may indicate only a local perspective on the relationship. Additional questions may be necessary to determine students' global understanding of this graph.

2. Describe what happens at $t = 1$.

    Learners may misinterpret what is happening at $t = 1$, supposing that the cars are at the same place. They may be thinking about distance rather than speed. Note that they may also indicate that the cars have the same speed. In such a case, an incorrect global interpretation of position in the graph overlays a correct local interpretation of values in the graph (speed). With guidance, learners can begin to dispel this misconception by moving back and forth between global and local perspectives.

3. Describe the relationship between the cars at $t = 0.25$ ($1/4$ hour).

   This question requires students to move between local and global perspectives on the graph. The graph indicates that at $t = 0.25$, car A is traveling 50 miles per hour and car B is moving at about 6 miles per hour. Thus, car A is traveling at a rate of speed that is more than 8 times that of car B. Challenge students to determine which car has traveled farther at this point and to explain how they know. Next, direct learners to describe what is happening at $t = 0.5$. Learners may also benefit from describing what the curved lines represent or explaining why one line curves up and the other curves down.

4. Which car travels farther?

   A key misconception is that a graph is a picture of the action in a story; the tendency to interpret a graph in this way may lead learners to suggest that the cars traveled the same distance and to justify that assertion by the height of the graph. This misconception can be addressed by guiding students to move between global and local perspectives on the graph, focusing alternately on the local values of the individual components of the graph (for example, the values and units of the axes and the points), and the global implications (the meaning of the curved lines, curving up versus curving down, and the covariation of the two variables).

5. What is the closest that the cars come to each other?

   Because the graph at $t = 1$ does not mean that the cars are at the same place, this question seems to be a logical one. The first step to take in estimating this relationship is to focus globally on the shape of the graphs of the two cars. First, car A has been traveling faster than car B for how long? (Approximately 1 hour.) Because car A has been traveling faster than car B for the entire hour, the closest the cars will be together is at $t = 0$. Note that students who answer this question correctly may nevertheless not understand the representation. For example, they may give the answer $t = 0$ because the graphs intersect at that value. In such cases, students may also suggest $t = 1$ as another possibility. Leading students through both local and global interpretations of the representation may help them develop a more robust understanding of such graphs.

   A global perspective might reveal that car A traveled over 50 miles per hour for $3/4$ hour, and car B traveled under 50 miles per hour for $3/4$ hour. Such a focus might also reveal that half the time car A was traveling at a speed greater than 75 miles per hour, whereas car B did not get to that speed until the final 8 minutes or so. The distance that the cars traveled can then be estimated (distance for car A ≈ 45 miles, distance for car B ≈ 35 miles). An interesting follow-up to this activity would be to plot the distance of the cars from the speed and time.

Two Walkers

The next example, also from Monk (2003), engages students in comparing increasing and decreasing rates. The example presents two walkers, Amanda and Joe, in the following situation:

> Amanda and Joe are standing next to each other and start walking at the same time, along parallel lines. Amanda starts by taking big steps, and each step grows smaller as she walks. Joe starts with small steps, and each step gets larger as he walks.

After giving your students the table in figure 4.5a, which shows Amanda's and Joe's steps and step sizes, ask them to describe the two walking trips. Before having the students make a graph of the trips, as shown in figure 4.5b, encourage them to act out the two trips.

	Step size (ft.)	
Step number(s)	Amanda	Joe
1	3.00	0.0
2	2.75	0.5
3	2.50	1.0
4	2.25	1.5
5	2.00	2.0
6	1.75	2.5
7	1.50	3.0
8	1.25	3.5
9	1.00	4.0
10	0.75	4.5

(a)

Fig. 4.5. Amanda's and Joe's step size at each step, shown in (a) a table and (b) a graph. From Monk (2003, p. 253).

Graphs as Representations of Functions

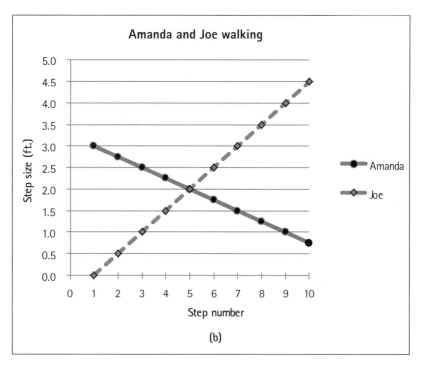

Fig. 4.5. Continued

This example is similar to that of the speed of two cars in that the unit rate for both graphs is changing, but it is different in that the unit rate for one individual is increasing while that for the other is decreasing. Introducing graphs of increasing and decreasing rates adds complexity, challenging students who have difficulty in constructing unit rates from graphs of linear functions. This example readily lends itself to an acting-out strategy: students may benefit from physically stepping out Amanda's and Joe's trips while their partners record and graph them and the subsequent results.

The new example offers learners another comparison of two objects moving at different rates of speed. This example not only slows down the action of the relationship between the two variables but also prompts learners to step through it. Learners should first graph the two trips. Then responding to a number of structured questions can help them focus on key components of the graph. The following questions are again interspersed with observations about students' likely responses:

1. What do Amanda's and Joe's trips look like? (global)

    Before graphing the values in the table, learners should attempt to predict the shape and general characteristics of the graph (global perspective). Because Amanda's steps are decreasing in size, some learners may think that the graph for Amanda's trip will curve down as the step number increases, whereas the graph for Joe's trip will curve up. (To clarify and extend such students' understanding, you might ask them to produce a situation where the graph would curve up or down.) Once students have made a graph showing Amanda's and Joe's trips, they should explain why it has the shape that it does, and what such a shape means. For example, why does each person's walking appear as a straight line? Why is one graph slanted up and the other slanted down, from left to right?

2. What happens at the fifth step, $s = 5$? (local)

The graphs of the two trips intersect at $s = 5$. It is important to focus students' attention on this fact and question them about it. What does this intersection mean? Who has traveled farther at this time? Who is traveling faster? In this case, global questions would focus on the general shape of the graph, why it has that shape, and what such an intersection means. Local questions would focus on the individual values of the points (noting, for example, that both travelers have the same step size at $s = 5$, so at that point they are traveling at the same speed), the intersection, and perhaps the slopes of the line.

3. What happens to Joe at $s = 1$? (local)

Ask about $s = 0$, since this is often a confusing point for students. Be sure to connect what Joe and Amanda are doing at $s = 0$ and how this is represented in the graph. What are Joe's and Amanda's positions at $s = 0$? Understanding $s = 0$ requires both a global and a local perspective. In this example, the steps are discrete. At $s = 0$, there are zero steps, so nothing has happened yet. Still, this knowledge is useful for understanding what happens at $s = 1$. Amanda's step is 3 feet. Joe's step is 0 feet. Locally, Joe's step is no different from his step at $s = 0$. Globally, however, his initial step has meaning beyond $s = 0$. How far apart are Joe and Amanda at $s = 0$? At $s = 1$?

4. Who travels farther? (local and global)

Some learners will automatically say that Joe travels farther because of the greater height of the graph that represents Joe's trip. This answer is correct—Joe does travel farther—but why? The height of the graph does not indicate distance but step number. Question 2 asked about the situation at $s = 5$. How far have Amanda and Joe traveled then? How far have Amanda and Joe traveled at $s = 10$?

5. Who is ahead at $s = 7$? (local)

   At step 7, Amanda is ahead. How far apart are Joe and Amanda at this point? Amanda has walked 15.75 feet, and Joe has walked 10.5 feet. Are they closer together or farther apart at step 8, as compared with step 7? Joe is catching up. At step 8, Amanda has walked 17 feet, and Joe has walked 14 feet.

6. At what time have Amanda and Joe traveled the same distance? Does Joe ever catch Amanda? (local and global)

   Learners can form small groups to act out the parts of Amanda and Joe. They can use tape on the floor, marked paper strips or lengths of rope, or other materials to indicate the step sizes. Learners might be surprised to find that they can model Joe's and Amanda's trips along two parallel paths. As two students play the parts of Joe and Amanda, another student can call off the step number. Students might also pay special attention to what happens at $s = 0$ and $s = 1$. Modeling this situation on a life-size graph is a powerful experience for some students, and the results may surprise many.

Working with this example may also help students distinguish among types of graphs. The graph in figure 4.6 provides a representation of the covariance of two variables (that is, distance vs. step), and the graph in figure 4.7 shows a representation that more closely illustrates the actual activity (that is, it more concretely depicts the actual steps that Amanda and Joe took). Each graph contains information about Amanda's and Joe's trips but depicts a different aspect of those trips. How are the illustrations in these graphs different from that in figure 4.5b? Do all these graphs contain the same information? Such questions not only help you diagnose the depth of your students' understanding of these graphs but also move students beyond considering the illustration as a picture. By thinking about and responding to the questions, they can engage in an examination of the local and global properties of the graph that can help them to grasp the underlying covariance of the variables in the relationship, thereby building a more robust understanding of the function.

Putting Essential Understanding of Functions into Practice in Grades 9–12

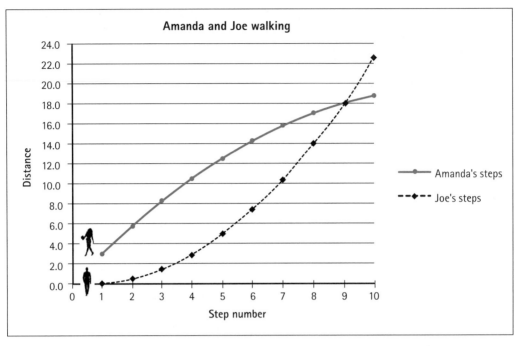

Fig. 4.6. A scatterplot of Amanda's and Joe's distance traveled by each step

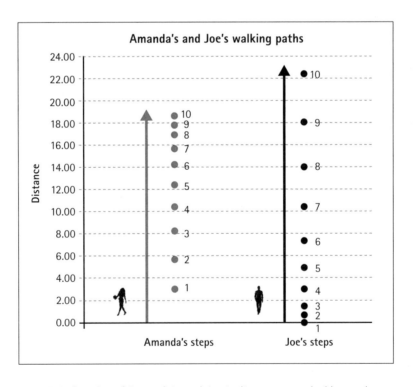

Fig. 4.7. A dot plot of Amanda's and Joe's distance traveled by each step

This activity offers an opportunity to pose additional questions to explore students' understanding of covariance. For example, if both Joe and Amanda take a step each second, the total time of their trips is 10 seconds. Who is traveling faster? When is Amanda traveling faster than Joe? When is Joe traveling faster than Amanda? Which of the three graphs (fig. 4.5, 4.6, or 4.7) provides the best help in answering these questions? What are the variables that represent the covariance for this set of questions? Reflect 4.3 offers an opportunity to think closely about the differences among the three graphs and the particular usefulness and value of each.

Reflect 4.3

Suppose that a student asks, "Which graph best helps answer the question about who was traveling faster?"

How might you respond?

After students have engaged in creating or examining a number of representations of the walking situation—a table of values, graphs (of step number vs. step size and step number vs. distance), and an illustration of Joe's and Amanda's trips—they should discuss the different information that these representations present, responding to questions such as the following:

- When and why would one representation be more useful than another?

- What insights do both local and global views provide?

- What other representations might provide additional information?

Assessing Understanding of Graphs

The examples that follow can help in assessing students' understanding of graphs, and they can be modified for use at multiple assessment intervals. Graphs in the first set appear in a story context that students can think about concretely as they make sense of the graphs and connect them with the story or a symbolic representation of a function. The second set of graphs leaves the context open so that students can create their own story situation to fit the graph and then consider a matching symbolic representation. In using the graphs with your students, be sure to question them from both local and global perspectives.

Putting Essential Understanding of Functions into Practice in Grades 9–12

The first example is taken from McKenzie and Padilla (1986; TOGS [Testing of Graphing in Science] assessment, questions 16 and 17) and offers an efficient way to assess or instruct students on the global properties of graphs. In this problem, set in the context of growing sunflowers, a student is investigating the effect of the size of the pot on the height of the plant. The graphs in figure 4.8 illustrate four possible outcomes. Students are provided with a description of a relationship, such as, "As the pot size increases, the plant size increases," and are asked to select the graph that matches it. Students might make additional global observations related to the shape and overall features of the graphs, remarking on slope, increasing and decreasing relationships, or the meaning of nonlinear relationships, for instance. These examples can be modified slightly to provide a set of experiences that address both global and local properties of graphs.

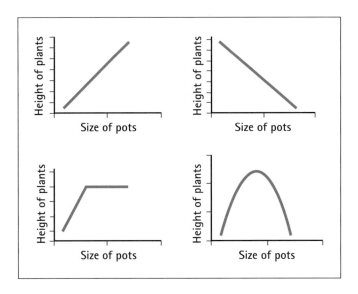

Fig. 4.8. Graphs of sunflower growth

Students can also be asked to work with a set of graphs such as those shown in figure 4.9. This example changes the assessment task. This time students need to come up with a situation that involves a relationship that could be represented by the graph. Students should identify the independent and dependent variables and discuss their relationship (covariance), and they might determine the meanings of the maximum and minimum of the graphs in the context of the situation (local and global). To assess your students' global interpretations, direct them to describe the meaning of the shape of each graph in the context of the situation.

Graphs as Representations of Functions

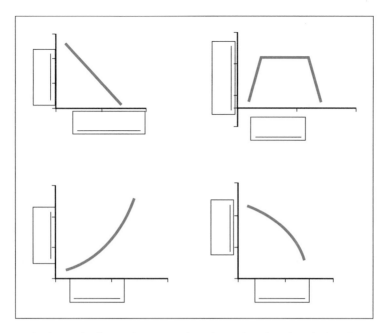

Fig. 4.9. Blank graphs for exploration of students' local and global understanding of graphs of functions

The graphs in figure 4.9 are sequenced to help move students from a pictorial view of a graph to an understanding of a graph as a dynamic representation of a situation. Graphs are not static pictures but representations of relationships. Without guided experiences that move learners between global and local perspectives on graphs, students may not develop the analytical tools necessary to interpret functions represented graphically. By emphasizing the importance of using both global and local lenses, you can help your students make the connections necessary to interpret graphs with meaning.

Conclusion

The next chapter—the last one in the book—invites you to take long looks back and ahead, at your students' earlier and later development of their understanding of functions, before and after grades 9–12. But before taking those long-distance views backward and forward, you may appreciate a recap—a brief review of the important understandings that you are charged with helping your students develop in their high school years, as discussed in Chapters 1–4.

Giving students a robust sense of function: Chapter 1

In the car-wash story that opens in Chapter 1 and weaves through the other chapters, Mr. Ramirez could easily have written on the board, "A function is a single-valued mapping of elements from one set to another." Derek and Marta could have written that down, and they might have even convinced themselves that they understood its meaning. But lost in that approach to functions would have been an appreciation of their utility and a recognition—and rejection—of a variety of misconceptions about them.

Instead, Mr. Ramirez presents functions as a practical answer to particular concrete needs. He does this overtly, by helping Derek and Marta use functions as tools for organizing a profitable fundraiser and by pointing out all the different occupations that make use of functions. He also does this more subtly, by engaging his students in activities that instill in them an awareness of the need for functions.

Mr. Ramirez provides function-related activities that immediately give his students a sense of confidence. They find it easy to walk to different positions along the wall in response to his instructions—at least at first. But then he complicates his instructions, turning them by just one degree and then one degree more, perturbing his students' confidence, puzzling them, challenging them. And then he teaches them a new word—*function*—whose meaning brings clarity and returns them to equilibrium.

Mr. Ramirez knows that it is insufficient to engage his students in a study of functions. Doing so is only part of his job. He is aware of many misconceptions that his students might develop in their studies, and he sets out to anticipate and confront each one directly. At times, that anticipation means that he keeps himself from teaching his students shortcuts and mnemonic devices, like the vertical line test, which will serve his students only in the short term, under limited circumstances. At other times, Mr. Ramirez's anticipation of misconceptions leads him to craft questions specifically to draw them out. The Function Finder activity includes examples designed to reveal whether or not students think that functions require every element in the range to have a single-valued mapping in the domain, for instance. If students think that a relationship is a function if and only if its inverse is also a function, Mr. Ramirez will discover that misconception early in their conceptual development.

Supporting emerging understanding of covariation: Chapter 2

Even though Derek and Marta work their way through several misconceptions and successfully learn the basic definition of a function, the particulars of that definition are still confusing to them. Covariation, in particular, is easily misunderstood or oversimplified, as demonstrated in Chapter 2, where Derek and Marta reappear in a vignette in which Derek attends to surface features in the range of a table (each of the entries is 5 soap bottles) while ignoring crucial changes in the domain. "Joint coordination"—the ability to hold two different ideas about the same object— is a prerequisite for a strong understanding of functions. Whether Derek and Marta are coordinating (a) changes in the domain and changes in the range, (b) a global and a local focus on a function, or (c) the abstract and concrete representations of a function, the need for joint coordination is ubiquitous, and adding this skill to your students' repertoires is well worth your investment of time and effort.

Your students' understanding of covariation may seem murky and difficult to assess, but Carlson's framework (2002) offers a roadmap. It is more than a robust tool for assessment. In fact, it is a hierarchical rubric that can help you point your students forward and upward. But even after you have identified students' ability, for example, to coordinate the *direction* of change (MA2), as well as their next goal of coordinating the *amount* of change (MA3), you may still find it difficult to chart a course for them from one level of covariational understanding to the next. Chapter 2 provides tasks, questions, and scenarios to help students make the transition to those higher levels in Carlson's framework.

Making combining, composing, and transforming functions natural activities: Chapter 3

Derek and Marta quickly understand that algebraic functions often represent a set of operations that they could perform on some inputs. For example, $f(x) = 2x + 3$ means taking the input, multiplying it by 2, and then adding 3. But as Chapter 3 illustrates through another vignette in the car-wash setting, the momentous realization for Marta and Derek is that *functions can be the inputs for other functions.* They can evaluate a function with another function just as though they are working with any real number. They can use a function as the input for *itself.* They can *combine* functions by using the sorts of operators they have previously used only on real numbers.

With that realization come more curiosity and questions about functions, and Mr. Ramirez guides the students in discovering new and powerful applications to facilitate their car wash. At the same time, the students' growing understanding brings with it a new set of misconceptions for Mr. Ramirez to anticipate and attend to. Domain and range, in particular, rush to the foreground after playing background roles during the students' development of their understanding of covariation. Before, they largely assumed that the domain and the range were real numbers or natural numbers. The domain could be counted on not to make trouble for the range. The graphs of those functions were free of holes and asymptotes. But when the students begin to combine, compose, and transform functions, they have to ensure that functions have the same domains. They have to declare, when dividing functions, which values can no longer be evaluated by the quotient.

Chapter 3 offers several examples of those combinations, compositions, and transformations, as well as two questions that are productive in the classroom:

1. What numbers no longer work?

2. Where do you find that number in the story?

The first question may serve to engage students. The ideas that numbers might not work in, or might in fact "break," a function can serve to humanize a discipline that often seems infallible. This question will give students the chance to play David to mathematics' Goliath as they look for the right numbers to bring a function low. The ideal outcome is that students will incorporate that question and others like it into their thought processes as they combine, compose, and transform functions in the future.

The second question, "Where do you find that number in the story?" challenges students who are skilled in symbolic manipulation to bear in mind both the *algebraic expression* that results from a combination and the *meaning* of that expression in the world. For example, in adding Derek and Marta's poster-making functions, a student may quickly and correctly combine like terms to get

$$(f+g)(x) = 5x - \frac{5}{6}.$$

But asking students what the 5 means in relation to Derek and Marta's story is an entirely different and more productive question. Remember, a concrete story is not just a useful enticement for students who feel alienated by abstract expressions. It also offers you a valuable check on your students' understanding.

Helping students interpret graphs of functions flexibly: Chapter 4

Chapter 4, the current chapter, demonstrates that students often reach for the most literal interpretation of stories represented in graphs. The opening vignette shows Marta's misinterpretation of Derek's graph of the numbers of cars washed. Marta's simple, understandable lapse is to assume that Derek's upward-trending line represents more productivity than her own downward-trending line. But a graph is not always a picture of a function. Unvaried experiences with height-time graphs (where the graph often *is* a picture of the context) can lead to these stereotypes and misconceptions.

Chapter 4 offers graphs for several scenarios, and in each case, the most literal interpretation of the direction of the graphs is also incorrect. In some cases, you could make a confident declaration of the sort that Derek makes to Marta and then ask your students to justify or disprove it. ("I think Derek was more productive. Am I right, or am I wrong? Tell your group what you think, and why you think so.") Here again, it is useful for students to attend to both global and local features of the graphs. To disprove your claim and develop a correct interpretation of the graph, students can ask themselves, "What do I think is happening two seconds into this scenario? Three seconds?" And then, advancing to a global vantage point, "What is the direction of the change here?"

Clearly, functions are more than arrows drawn from one set to another, and more than vertical lines slicing through a graph. Understanding functions requires abundant mental resources and the coordination of several different ideas at once. Work with functions provides calisthenics for the brain and career training for future statisticians, Web developers, and workers in many other fields.

Supporting meaningful teaching of functions is well worth the time that we have taken to write this volume, and we believe that developing students' understanding of functions is well worth a large share of your limited classroom time. Even though the restrictions imposed by the clock may encourage you to teach definitions without context, mnemonics without understanding, and examples without appreciation for the misconceptions that they might create, we urge you to take time to pose provocative questions about productive contexts, a sampling of which we have provided. Your students may struggle with the work, but the results will more than repay your efforts—if not immediately, on in-class assessments, then years later, when your students run a successful and wildly profitable car wash or another project to benefit a great cause.

into practice

Chapter 5
Looking Back and Ahead with Functions

This chapter—the last in this book—looks in opposite directions to frame the discussion of developing students' essential understanding of functions in grades 9–12. First, it looks back at the presecondary years to survey the work that helps students lay the foundation for understanding and using functions in grades 9–12. Then it looks ahead, surveying the ways in which students can continue to use and extend their understanding of functions in the postsecondary years. These two perspectives set the work that students need to do with functions in grades 9–12 in a broad context that underscores both its roots and its continuing growth and value.

Building a Base for Functions in Kindergarten–Grade 8

Functions are treated primarily informally in kindergarten–grade 7. Figure 5.1 shows a function-related standard for grade 6 included in the Common Core State Standards for Mathematics (CCSSM; National Governors Association Center for Best Practices and Council of Chief State School Officers [NGA Center and CCSSO] 2010). However, CCSSM presents this standard in the domain "Expressions and Equations" rather than explicitly under a "Functions" heading.

This grade 6 standard clearly addresses function-like characteristics and encourages students to begin thinking about functional relationships. To meet this standard, students must think about how two quantities covary, how one quantity is an "input" and the other is an "output," and how functions can be represented graphically, numerically, and analytically.

Putting Essential Understanding of Functions into Practice in Grades 9–12

Common Core State Standards for Mathematics, Grade 6

Represent and analyze quantitative relationships between dependent and independent variables.

9. Use variables to represent two quantities in a real-world problem that change in relationship to one another; write an equation to express one quantity, thought of as the dependent variable, in terms of the other quantity, thought of as the independent variable. Analyze the relationship between the dependent and independent variables using graphs and tables, and relate these to the equation.

Fig. 5.1. Expressions and Equations, CCSSM 6.EE.9 (NGA Center and CCSSO, p. 44)

Sixth-grade students can discover and investigate these ideas in a variety of concrete situations. For example, they should explore situations similar to the following:

> Donna is making a wall chart to list the heights of her family members. She makes a reference mark every 12 inches and labels the mark in both feet and inches. She has generated a table of feet and inches for her labels (shown in fig. 5.2).

f: number of feet	0	1	2	3	4	5	6	7
i: number of inches	0	12	24	36	48	60	72	84

Fig. 5.2. Donna's table of feet and inches

Donna explains to her younger brother Sam that this is an example of a *function* because if she is given a certain number of feet, she can find the corresponding number of inches by using the formula $i = 12f$. Sam asks, "Is it also a function because if you are given a certain number of inches, you can find the corresponding number of feet?"

Pause to examine Sam's question, guided by Reflect 5.1.

Reflect 5.1

What is Sam trying to clarify for himself with his question, "Is [the relationship between feet and inches] also a function because if you are given a certain number of inches, you can find the corresponding number of feet?"

How do you think Donna might respond to Sam's question?

At this level, the important function-related concepts are the following:

1. A function is a relationship between two sets, the domain and the co-domain, such that every element of the first set is paired with a unique element of the second set.

2. The range is a subset of the co-domain that contains the elements that are paired with elements from the domain.

3. A table of values (such as that in fig. 5.2) can be made to represent the function, since every "input" (here, number of feet) has a unique "output" (number of inches).

4. The relationship between inputs and outputs can be generalized by a formula (just as the relationship between feet and inches can be represented by $i = 12f$, for nonnegative values of f).

In this example, the domain is the set of whole numbers. Different sets could be chosen for the co-domain, with one also being the set of whole numbers. The range is the subset of the whole numbers that actually appear in the table—namely, the multiples of 12 up to 84. However, Sam asks the interesting question, which comes down to, "How do you know which quantity is the input and which one is the output?" Answering this question helps solidify understanding of the role of a function's domain and co-domain and paves the way for exploring a function's inverse (if it exists).

As the ideas in previous chapters suggest, the understanding that sixth graders must have to meet this standard lays the foundation for the conceptual understanding of functions that students need for working with functions at the secondary school level and beyond. For example, students are likely to struggle if they do not completely comprehend the example in figure 5.3, where the function $f(x) = x^2 + 2x - 3$ is represented by a graph and a table of sample values from the domain and co-domain, in addition to its equation.

Presecondary students should have experience in representing functions in a variety of ways that is sufficient to enable them to choose with ease the representation that best fits a particular situation. Teachers must also ensure that students understand the difference between the concept of function and various ways to represent a function.

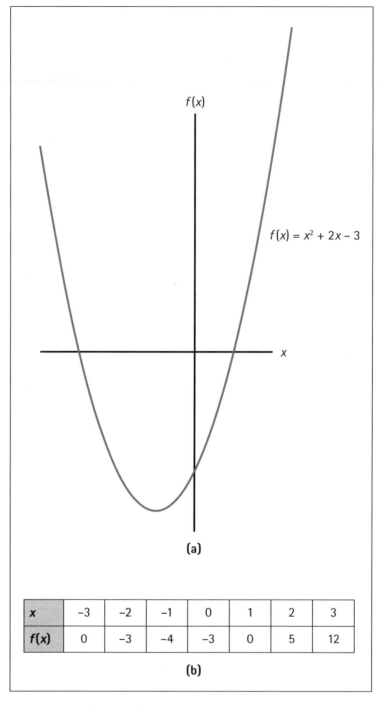

Fig. 5.3. The function $f(x) = x^2 + 2x - 3$, represented by (a) a graph and (b) a table of values

In CCSSM, the topic of functions appears formally for the first time at grade 8, where five standards are sorted into two clusters in a new domain, "Functions," which then plays a leading role as one of the six conceptual categories of high school mathematics, along with "Number and Quantity," "Algebra," "Modeling," "Geometry," and "Statistics and Probability" (NGA Center and CCSSO 2010). Figure 5.4 shows the function standards that CCSSM identifies for grade 8.

Common Core State Standards for Mathematics, Grade 8

Define, evaluate, and compare functions.

1. Understand that a function is a rule that assigns to each input exactly one output. The graph of a function is the set of ordered pairs consisting of an input and the corresponding output.

2. Compare properties of two functions each represented in a different way (algebraically, graphically, numerically in tables, or by verbal descriptions). *For example, given a linear function represented by a table of values and a linear function represented by an algebraic expression, determine which function has the greater rate of change.*

3. Interpret the equation $y = mx + b$ as defining a linear function, whose graph is a straight line; give examples of functions that are not linear. *For example, the function $A = s^2$ giving the area of a square as a function of its side length is not linear, because its graph contains the points (1, 1), (2, 4), and (3, 9), which are not on a straight line.*

Use functions to model relationships between quantities.

4. Construct a function to model a linear relationship between two quantities. Determine the rate of change and initial value of the function from a description of a relationship or from two (x, y) values, including reading these from a table or from a graph. Interpret the rate of change and initial value of a linear function in terms of the situation it models, and in terms of its graph or a table of values.

5. Describe qualitatively the functional relationship between two quantities by analyzing a graph (e.g., where the function is increasing or decreasing, linear or nonlinear). Sketch a graph that exhibits the qualitative features of a function that has been described verbally.

Fig. 5.4. Functions, CCSSM 8.F.1–5 (NGA Center and CCSSO 2010, p. 55)

The first two standards, 8.F.1 and 8.F.2, echo the previous discussion, whereas the remaining standards, 8.F.3–8.F.5, illustrate the critical role that an early understanding of function can play in helping students be more successful in the secondary grades.

For instance, consider the following example:

> A taxi company advertises a rate of $1.50 plus $0.25 for each $1/10$ mile. What is the fare for a trip that is 3 miles long?

This simple example has many important ideas associated with it. First, this situation does describe a function, with the number of miles serving as the independent variable and the fare serving as the dependent variable. Furthermore, the domain consists of all positive real numbers (probably rounded to the nearest tenth of a mile), one possible co-domain consists of all positive real numbers greater than or equal to $1.50, and the range is the set of actual fares.

Second, students can engage in a rich discussion about how to represent this function properly. Some students may argue for using a linear function and for representing the function analytically as $f = 1.5 + 0.25(10m) = 1.50 + 2.5m$, where m is the number of miles traveled (rounded down to the nearest $1/10$ mile), and f is the corresponding fare (see fig. 5.5).

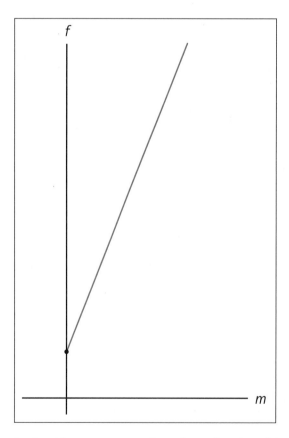

Fig. 5.5. A graph of taxi fares represented as a linear function, $f(x) = 1.50 + 2.5m$

Other students may argue for a step-wise linear function, since the taxi fare will be $1.50 for any trip that is between 0 miles and $1/10$ mile, $1.75 for any trip that is greater than or equal to $1/10$ mile but less than $2/10$ mile, and so on (see fig. 5.6). Reflect 5.2 offers an opportunity to consider how to discuss the differences between the graphs in figures 5.5 and 5.6 with students.

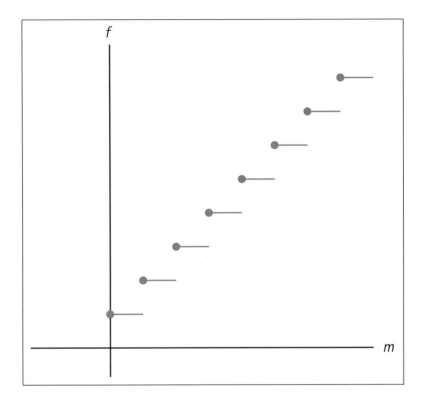

Fig. 5.6. A graph of taxi fares represented as a piecewise linear function

Reflect 5.2

Figures 5.5 and 5.6 show two different graphs that represent the fares of a taxi company whose rate is $1.50 plus $0.25 for each $1/10$ mile.

As a teacher, what could you do to help students interpret and understand the differences between the graphs?

Several questions are useful to ask students in relation to the graphs:

1. Is the $0.25 rate applied uniformly throughout the distance interval that is rounded down to $1/10$ mile, or is it applied only at the end of each $1/10$-mile segment?

2. Why does the taxi company advertise $0.25 for each $1/10$ mile instead of $2.50 for each mile?

3. What are the implications, if any, of choosing one model over the other?

Students may also debate the type of representation that should be used in this situation. Does a formula have inherent difficulties or limitations? Does a table? Is a graph an easier or more efficient way to represent the function?

As another example, consider the situation of a skydiver who jumps out of a plane at an altitude of 6000 meters. After the first 10 seconds, he falls at a constant rate. The graph in figure 5.7 shows his altitude at 10-second intervals.

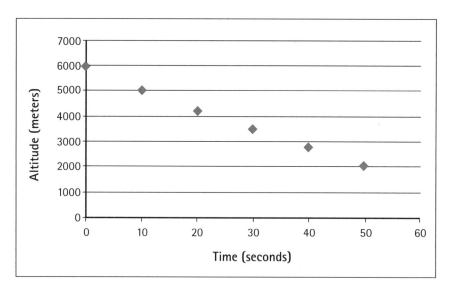

Fig. 5.7. A graph of skydiver's height as a function of time

After the first 10 seconds, the equation describing the skydiver's altitude a as a function of time t is $a = 5000 - 750t$. Standard 8.F.4 recommends that students explore graphs such as this. They can interpret the graph to answer questions like the following:

1. What do the values 5000 and −750 in the equation represent in the skydiver's situation?

2. What does the *y*-intercept of 6000 represent in the skydiver's situation?

3. Why might the change in altitude in the first 10 seconds (1000 meters) be different from the constant change in altitude (750 meters) in subsequent 10-second intervals?

4. How would the function change if the –750 in the equation were changed to –650 (or to –850)? How would the graph change, and what would that mean in the story?

It is very important to give students numerous early opportunities to explore, experience, and discuss all aspects of functions in a wide variety of representations. Furthermore, these experiences should include a varied sampling of the many real-life examples of functions that exist. Students who experience functions only as abstract "input-output" machines, or only as graphs, or only as tables of values will have difficulty later in appreciating the rich variety of representations that are possible for functions. They should be able to predict the impact of a change in a component of one representation on another representation.

For example, in the skydiving example, students should be able to determine how changing –750 to –650 in the equation $a = 5000 - 750t$ affects the graphical and numerical representation of the situation in a predictable way (that is, the slope of the linear part of the graph changes from –750 to –650) and should understand that in the function's table of values each 10-second change in time corresponds to a –650-meter change in altitude). Teachers must give students a chance to correct misconceptions and build the deep conceptual understanding that is necessary for later mathematical success.

Extending Understanding of Functions after Grades 9–12

Functions also play an important role in postsecondary school mathematics. Therefore, students must build a robust conceptual understanding of the ideas presented in previous chapters to be successful in their continued studies of mathematics. In addition to understanding what a function is, more advanced students are usually required to determine certain characteristics of a given function, such as whether it is *onto, one-to-one, continuous,* or *bounded.*

Consider, for example, a function that models a situation in which six people are assigned to inspect four restaurants, and, as a quality-control measure, two of the restaurants are chosen for inspection by two different people. Figure 5.8 represents the situation.

Putting Essential Understanding of Functions into Practice in Grades 9–12

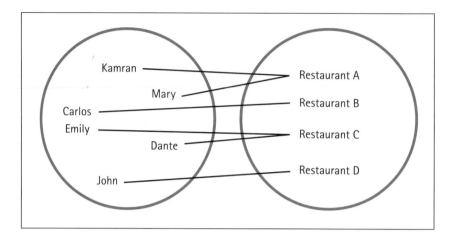

Fig. 5.8. Six people assigned to inspect four restaurants, with two restaurants checked by two different inspectors

This is an *onto* function: in the co-domain, every element (a restaurant) is visited by at least one inspector—no restaurant is left out. Another way to confirm that this is an onto function is to note that the co-domain and the range are the same set. However, the function is not *one-to-one*, since some restaurants—specifically, A and C—are visited by more than one inspector—Kamran and Mary for restaurant A, and Emily and Dante for restaurant C. In a one-to-one function, not only is each element of the domain paired with a unique element of the range, but each element of the range is also paired with a unique element of the domain. We would have a one-to-one (and onto) function if the inspection did not require the quality-control measure.

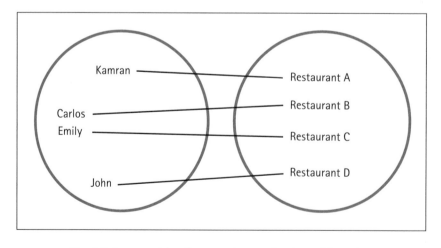

Fig. 5.9. An example of an one-to-one (and onto) function

Students may wonder if it is possible for a function to be one-to-one but not onto. The answer is yes, as figure 5.10 illustrates.

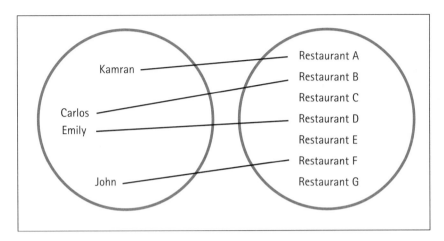

Fig. 5.10. An example of a one-to-one (but not onto) function

It may be that not every restaurant is inspected every year, in which case the figure might represent the assignment of inspections for the current year, with restaurants C, E, and G unscheduled in this cycle of inspections. Of course, any function can be made to be an onto function simply by restricting the co-domain to those elements in the range.

Students studying higher-level mathematics often deal with increasingly more complicated functions, multi-valued functions, and functions that cannot be represented by a graph in the Cartesian plane. Hence, it is important that the misconceptions previously discussed are remedied before students leave secondary school.

Postsecondary school mathematics expands the number, type, and complexity of functions that students may encounter. For example, in calculus, the antiderivative function $F(x)$ of another function $f(x)$ is often of interest, e.g.,

$$F(x) = \int_0^x f(t)\,dt.$$

As pictured in figure 5.11, a particular value, $F(r)$, of this function can be thought of as the area between the x-axis and the graph of $f(x)$ between the values 0 and r. What is important is the complexity of the idea that the function $F(x)$ is defined in terms of another function, $f(x)$.

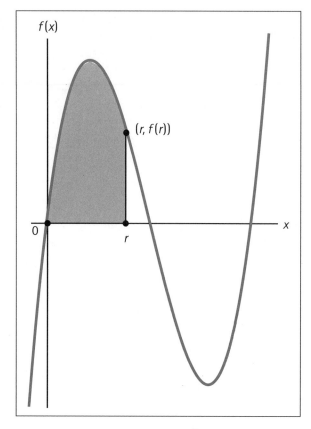

Fig. 5.11. Illustration of $F(x) = \int_0^x f(t)\,dt$ as an area

Or consider the function $f(t) = 49\left(1 - e^{-0.2t}\right)$, which is graphed in figure 5.12 and represents the velocity of a falling object, where $t \geq 0$ is measured in seconds and $f(t)$ is measured in meters per second.

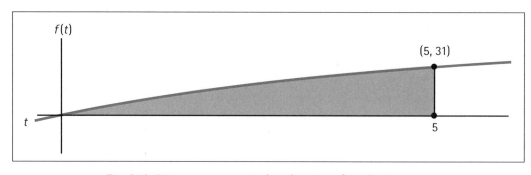

Fig. 5.12. Distance represented as the area of a velocity curve

The total distance that the object has fallen after x seconds have elapsed is represented by the function

$$F(x) = \int_0^x f(t)\,dt.$$

The point (5, 31) tells us that after 5 seconds the object is falling at a rate of (approximately) 31 meters per second, and the shaded area under the graph of $f(t)$ represents the total distance that the object has fallen (approximately 90 meters). Whether students understand the notion of a function being *continuous* or *bounded* is relevant. In an inexact way, it is possible to say that the function $f(t)$ is continuous since its graph can be drawn without lifting a pencil. Although it is not as easy to see, it is also possible to say that $F(x)$ is continuous since the object cannot travel from a distance of 20 meters, for example, to a distance of 25 meters without traveling all the distances between 20 and 25 meters.

Although it is not necessarily apparent from the graph, the function $f(t)$ is bounded, below by 0 (the initial velocity) and above by 49 (the terminal velocity); that is, $0 \le f(t) \le 49$. What about $F(x)$? In general, $F(x)$ is bounded below by 0 (the initial distance traveled) but not bounded above. However, in practice, $F(x)$ is bounded above by its initial height: the object cannot fall any farther once it hits the ground.

The previous example also demonstrates how important it is that students have experiences in representing functions in multiple ways. Figure 5.13 shows a partial table of values for t and $f(t)$.

t	0	1	2	3	4	5	6	7	8
$f(t)$	0	8.9	16.2	22.1	27.0	31.0	34.2	36.9	39.1

Fig. 5.13. A table of values for $f(t) = 49(1 - e^{-0.2t})$ for selected values of t

There is nothing wrong with this numerical representation of the function. The approximate distance that the object falls (and hence, the approximate value of $F(x)$) can be computed from the table. However, the numerical representation simply does not have the same impact as the graph.

Another example of such a function comes from mathematical statistics, where the probability that a random variable X has a value less than or equal to b is given by

$$F(b) = P(X \leq b) = \int_{-\infty}^{b} f(t)\,dt.$$

The function $f(t)$ is called the *probability density function*, and $F(b)$ is represented by the shaded area under the graph of $f(t)$ in figure 5.14, which shows an example where X is a normally distributed random variable. The function $F(b)$ is commonly called a *cumulative distribution function*.

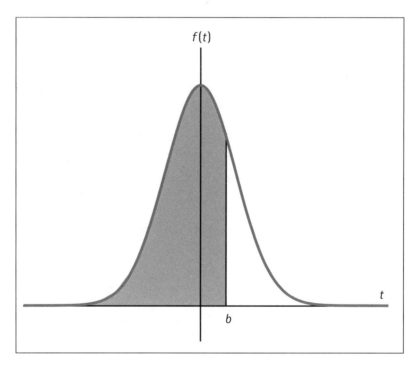

Fig. 5.14. A cumulative distribution function as the area under the graph of its associated probability density function

Probability density functions have numerous applications, one of which is in an aspect of project management called *program evaluation and review technique* (PERT). If, for example, in the construction of a single-family house, the project manager knows that laying the house's foundation will take between two and five days, he can use the probability density function $f(t)$ graphed in figure 5.15 to model the time X (in days) necessary for the task.

Looking Back and Ahead with Functions

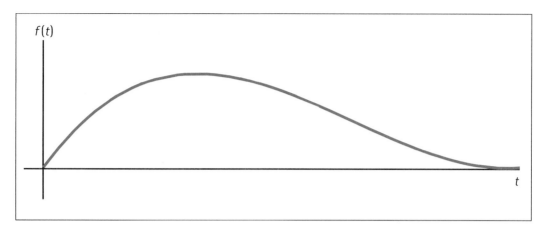

Fig. 5.15. A graph of probability density function for laying the foundation of a house

To schedule subsequent work, the project manager must make a decision about how long it will take to finish the foundation. PERT theory tells him the probability that laying the foundation will take at most three days. The shaded area under the graph in figure 5.16 represents the probability, which is .407, given by

$$P(X \leq 3) = \int_0^I f(t)\,dt \approx 0.407.$$

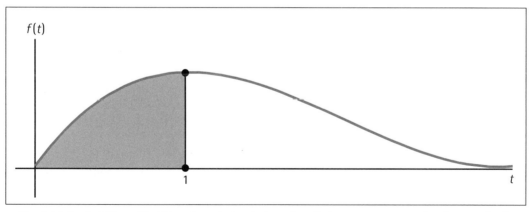

Fig. 5.16. Probability of finishing foundation in less than 3 days as an area under the graph of a probability density function

As mentioned earlier, the point we are emphasizing is that students must have ample opportunity to develop a conceptual understanding of functions that will

Putting Essential Understanding of Functions into Practice in Grades 9–12

allow them to be able to tackle problems involving more complicated instances of functions, such as cumulative probability functions, which are defined in terms of probability density functions.

Whereas the graphs of two-dimensional curves are commonly studied in secondary school, the surfaces of three-dimensional curves are often explored in higher mathematics. For example, the graph in figure 5.17 gives the concentration $C = f(x, t)$ of a drug in the bloodstream as a function of the amount x of the drug injected and the time t, measured in hours since its injection. In the graph in figure 5.17, C is measured in milligrams per 100 milliliters of blood, x is measured in hundreds of milligrams, and t is measured in hours. This graph can be described by the equation

$$C = f(x, t) = te^{-t(5-x)}.$$

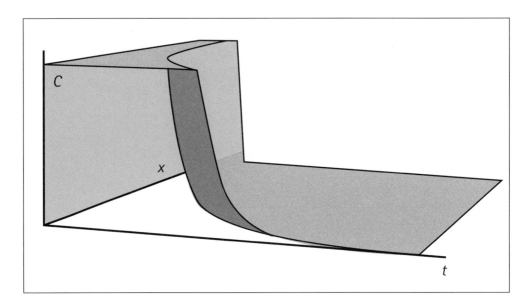

Fig. 5.17. A graph of $C = f(x, t) = te^{-t(5-x)}$

In this case, there are two independent variables, x and t, and one dependent variable, C. Thus, the domain is the set of ordered pairs, (x, t), where $0 \leq x \leq 4$ and $t \geq 0$. This ordered pair is matched with a number C from the co-domain, where $C \geq 0$. The range is the set of all the values of C that actually are matched with an element from the domain. (Potentially, $C \geq 0$, but in actuality, C is usually less than 0.4.) For example, if 300 milligrams of the drug are injected ($x = 3$), then after 2 hours ($t = 2$), the concentration of the drug in the blood stream is $2e^{-4}$, or approximately 0.0367 milligrams per 100 milliliters.

This point can be found on the surface above by moving 3 units in the x-direction, 2 units in the t-direction, and 0.0367 units in the C-direction. Functions of more than one variable can be quite sophisticated, and it should be evident how important it is for students to develop a deep conceptual understanding of functions and their graphs—far beyond an ability to apply the so-called vertical line test—before they study higher-level mathematics. With this in mind, stop to consider the question in Reflect 5.3.

Reflect 5.3

Suppose that you overheard one of your students say, "A relationship cannot be a function if it does not pass the vertical line test."

How would you respond?

Some students over-contextualize a function's graph and define concepts related to a function in terms of its graph. For example, in responding to the graph in figure 5.18, some students might claim that the *graph* is a function because it passes the vertical line test. That is, they might suppose that the *graph itself* is a function because it intersects any vertical line at only one point. Furthermore, they might claim that the *graph* does not have an inverse function because it does not pass the horizontal line test: at least one horizontal line intersects the graph in more than one point. Using this thinking, students might believe that a function is "any graph that passes the vertical line test," and a function has an inverse if "the graph passes the horizontal line test."

To examine another misconception, consider the experiment in which a fair coin is flipped five times. If X is the random variable representing the number of times that heads occurs in these five flips, then the probability distribution for X is that shown in figure 5.19, first in tabular and then in graphical form.

Do the vertical line test and the horizontal line test apply to the graph of a discrete function? Some students might think that they do not. Because of a weak understanding of a function's domain, these students might believe that the points must be connected in a curve, as in figure 5.20, for the graph "to be a function," permitting the use of the horizontal and vertical line tests. This common misconception often prevents students from developing the conceptual understanding of functions that is necessary for mastering higher-level mathematics. Reflect 5.4 poses a further question with implications for students who identify the graph as the function as determined by the vertical line test.

Putting Essential Understanding of Functions into Practice in Grades 9–12

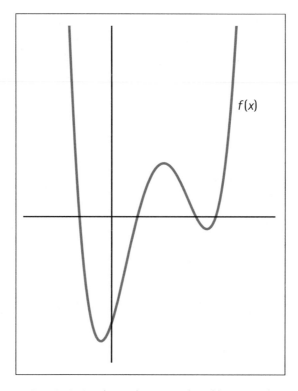

Fig. 5.18. Applying the vertical and horizontal lines tests to the graph of a function

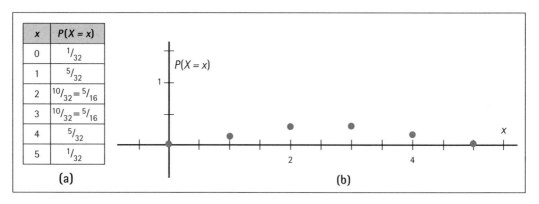

Fig. 5.19. The probability distribution function for the discreet random variable X, shown in (a) a table and (b) a graph

Looking Back and Ahead with Functions

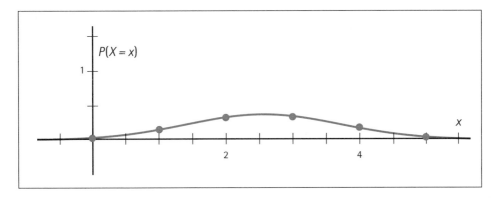

Fig. 5.20. The probability distribution function for a continuous random variable X

Reflect 5.4

Are there functions that cannot be graphed?

Besides the conceptual difficulty involved in thinking about the graph as the function instead of as a *representation* of the function, an additional problem arises from associating functions too closely with graphs and the vertical line test. How does a student with these misconceptions interpret function concepts if a function cannot be represented by a graph? For example, in figure 5.21, triangle *ABC* is reflected across line *l* to form triangle *A'B'C'*. Reflecting the entire plane across line *l* is an example of a function, since each point in the plane (the domain) gets reflected to one and only one point in the plane (both the co-domain and the range). Furthermore, this function has an inverse—namely, itself, a reflection across line *l*.

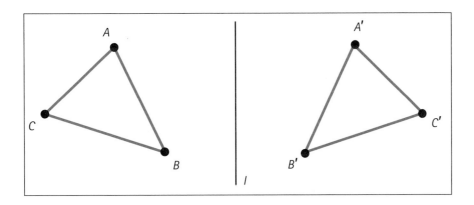

Fig. 5.21. The reflection of triangle *ABC* across line *l*

We know that this function has an inverse because a reflection in the plane is a one-to-one and onto relationship, and it is precisely functions that are both one-to-one and onto that have inverses. However, this function cannot be "graphed." Students' misconceptions about functions may cause them to be confused about how the vertical line test and the horizontal line test could be applied to determine whether a reflection is a function, and if it is, whether it has an inverse.

Transformations are studied extensively in higher mathematics, and since transformations are functions, various operations can be performed on them. However, contrary to some students' beliefs, some rules that they learn to apply on numerical functions do not apply to all functions. For example, secondary school students often explore how the graph of $f(x)$ is related to the graphs of $f(x + h)$, $f(x) + h$, and $af(x)$, for constants h and a. However, these ideas have no applicability to functions in general, particularly those whose domains are numerical or those that cannot be represented by graphs, like transformations.

Moreover, combining two functions into one through function composition is an operation that does have universal applicability and one that is used frequently in higher mathematics. Isometries (such as translations, rotations, reflections, and glide reflections) are transformations that preserve distances, and they are an important class of geometric functions. It can be shown that the composition of two isometries is itself an isometry. For example, if σ_r and σ_s are functions representing reflections across the lines r and s, respectively, then $(\sigma_s \circ \sigma_r)(X) = \sigma_s[\sigma_r(X)]$ takes the set of points X and maps it to its image by first reflecting X across line r to get $\sigma_r(X)$ and then reflecting $\sigma_r(X)$ across line s, as shown in figure 5.22, where triangle ABC is mapped to triangle $A'B'C'$ by successive reflections across lines r and s.

Also important to the study of transformations is determining the single transformation that is equivalent to the composition (or product) of two given transformations. In the current example, a reflection across line r followed by a reflection across line s is equivalent to a rotation of 2α degrees around point P, where α is the angle at which lines r and s intersect, and P is the point at which the two lines intersect, as illustrated in figure 5.23.

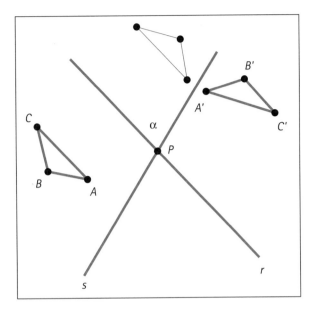

Fig. 5.22. Successive reflections across lines r and s

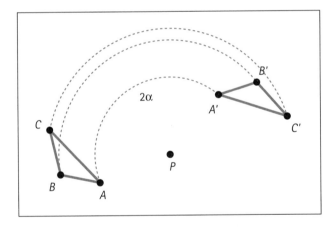

Fig. 5.23. The composition of the reflections in figure 5.22 is equivalent to a rotation of 2α degrees about point P.

Similarly, if the two lines r and s are parallel, as in figure 5.24, then successive reflections across lines r and s are equivalent to a different transformation—a translation, which can be represented by the vector $\vec{v}$, as shown in figure 5.25. This translation, $\vec{v}$, is congruent to $\overrightarrow{AA'}$, $\overrightarrow{BB'}$ and $\overrightarrow{CC'}$. Its length is equal to twice the distance between lines r and s, and its direction is perpendicular to both r and s.

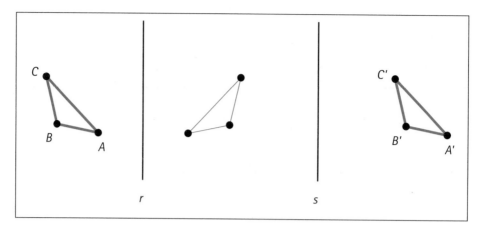

Fig. 5.24. The composition of successive reflections across parallel lines *r* and *s*

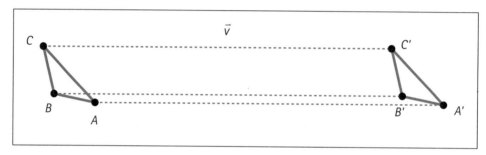

Fig. 5.25. The composition of the reflections in figure 5.24 is equivalent to a translation, $\vec{v}$

Of frequent interest in the study of transformations in higher mathematics is the operation that can be thought of as the "reverse" of composition: factoring. It is known that any isometry can be written as the product of at most three isometries. For example, a glide reflection is defined as the product of a reflection and a translation in the direction parallel to the line of reflection. As illustrated in figure 5.26, the point *A* is mapped to the point *A'* by first reflecting it across the line *t* and then translating it by the vector $\vec{v} = \overrightarrow{QR}$.

However, since a translation can be written as the product of successive reflections across parallel lines, a glide reflection can be written as the product of three reflections. That is to say, the glide reflection described above is equivalent to successive reflections across lines *t*, *s*, and *r*, where *r* and *s* are parallel to each other and perpendicular to *t*. Furthermore, the distance between lines *r* and *s* is one-half the distance *QR*.

Looking Back and Ahead with Functions

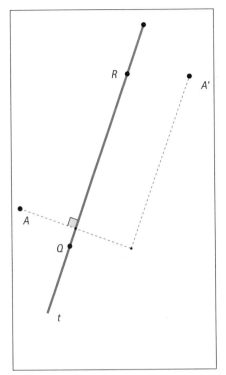

Fig. 5.26. A glide reflection is the composition of a reflection and a translation.

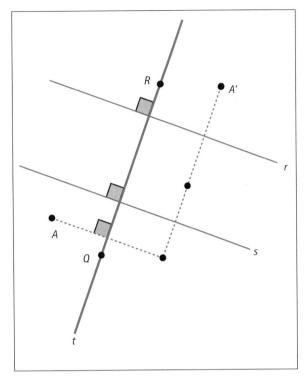

Fig. 5.27. A glide reflection as the product of three reflections

Conclusion

At the beginning of this chapter, we presented a scenario in which Donna is trying to explain what a function is to her younger brother, Sam. As you can see, a complete understanding of function is much more complicated than just substituting a value into an equation and computing the result. It requires knowledge of domains, co-domains, and ranges; skill in making sense of multiple representations of functions; and understanding of the ways in which functions model real-world phenomena and the variables in a function change in relationship to each other in a predictable way (covariance). Teachers have a tremendous responsibility to give students high-quality opportunities to develop the deep conceptual knowledge of functions that will prepare them to be successful in their study of mathematics, regardless of the level to which they progress.

Appendix 1
The Big Ideas and Essential Understandings for Functions

This book focuses on essential understandings that are identified and discussed in *Developing Essential Understanding of Functions for Teaching Mathematics in Grades 9–12* (Cooney, Beckmann, and Lloyd 2010). For the reader's convenience, the full list of the big ideas and essential understandings in that book is reproduced below. The essential understandings that are the special focus of this book are highlighted in avocado.

Big Idea 1. *The function concept.* The concept of function is intentionally broad and flexible, allowing it to apply to a wide range of situations. The notion of function encompasses many types of mathematical entities in addition to "classical" functions that describe quantities that vary continuously. For example, matrices and arithmetic and geometric sequences can be viewed as functions.

Essential Understanding 1*a*. Functions are single-valued mappings from one set—the *domain* of the function—to another—its *range*.

Essential Understanding 1*b*. Functions apply to a wide range of situations. They do not have to be described by any specific expressions or follow a regular pattern. They apply to cases other than those of "continuous variation." For example, sequences are functions.

Essential Understanding 1*c*. The domain and range of functions do not have to be numbers. For example, 2-by-2 matrices can be viewed as representing functions whose domain and range are a two-dimensional vector space.

Big Idea 2. *Covariation and rate of change.* Functions provide a means to describe how related quantities vary together. We can classify, predict, and characterize various kinds of relationships by attending to the rate at which one quantity varies with respect to the other.

Essential Understanding 2*a*. For functions that map real numbers to real numbers, certain patterns of covariation, or patterns in how two variables change together, indicate membership in a particular family of functions and determine the type of formula that the function has.

Essential Understanding 2*b*. A rate of change describes how one variable quantity changes with respect to another—in other words, a rate of change describes the covariation between two variables.

Essential Understanding 2*c*. A function's rate of change is one of the main characteristics that determine what kinds of real-world phenomena the function can model.

Big Idea 3. *Families of functions.* Functions can be classified into different families of functions, each with its own unique characteristics. Different families can be used to model different real-world phenomena.

Essential Understanding 3*a*. Members of a family of functions share the same type of rate of change. This characteristic rate of change determines the kinds of real-world phenomena that the functions in the family can model.

Essential Understanding 3*b*. Linear functions are characterized by a constant rate of change. Reasoning about the similarity of "slope triangles" allows deducing that linear functions have a constant rate of change and a formula of the type $f(x) = mx + b$ for constants m and b.

Essential Understanding 3*c*. Quadratic functions are characterized by a linear rate of change, so the rate of change of the rate of change (the second derivative) of a quadratic function is constant. Reasoning about the vertex form of a quadratic allows deducing that a quadratic has a maximum or minimum value and that if the zeros of the quadratic are real, they are symmetric about the *x*-coordinate of the maximum or minimum point.

Essential Understanding 3*d*. Exponential functions are characterized by a rate of change that is proportional to the value of the function. It is a property of exponential functions that whenever the input is increased by 1 unit, the output is multiplied by a constant factor. Exponential functions connect multiplication to addition through the equation $a^{b+c} = (a^b)(a^c)$.

Essential Understanding 3*e*. Trigonometric functions are natural and fundamental examples of periodic functions. For angles between 0 and 90 degrees, the trigonometric functions can be defined as the ratios of side lengths in right triangles; these functions are well defined because the ratios of side lengths are equivalent in similar triangles. For general angles, the sine and cosine functions can be viewed as the y- and x-coordinates of points on circles or as the projection of circular motion onto the y- and x-axes.

Essential Understanding 3f. Arithmetic sequences can be thought of as linear functions whose domains are the positive integers.

Essential Understanding 3g. Geometric sequences can be thought of as exponential functions whose domains are the positive integers.

Big Idea 4. *Combining and transforming functions.* Functions can be combined by adding, subtracting, multiplying, dividing, and composing them. Functions sometimes have inverses. Functions can often be analyzed by viewing them as made from other functions.

Essential Understanding 4a. Functions that have the same domain and that map to the real numbers can be added, subtracted, multiplied, or divided (which may change the domain).

Essential Understanding 4b. Under appropriate conditions, functions can be composed.

Essential Understanding 4c. For functions that map the real numbers to the real numbers, composing a function with "shifting" or "scaling" functions changes the formula and graph of the function in readily predictable ways.

Essential Understanding 4d. Under appropriate conditions, functions have inverses. The logarithmic functions are the inverses of the exponential functions. The square root function is the inverse of the squaring function.

Big Idea 5. *Multiple representations of functions.* Functions can be represented in multiple ways, including algebraic (symbolic), graphical, verbal, and tabular representations. Links among these different representations are important to studying relationships and change.

Essential Understanding 5a. Functions can be represented in various ways, including algebraic means (e.g., equations), graphs, word descriptions, and tables.

Essential Understanding 5b. Changing the way a function is represented (e.g., algebraically, with a graph, in words, or with a table) does not change the function, although different representations highlight different characteristics, and some may show only part of the function.

Essential Understanding 5c. Some representations of a function may be more useful than others, depending on the context.

Essential Understanding 5d. Links between algebraic and graphical representations of functions are especially important in studying relationships and change.

Appendix 2
Resources for Teachers

The following list highlights a few of the many websites that offer helpful resources for teaching functions in grades 9–12.

Technology

Core Math Tools
The NCTM website contains a wide variety of technology software called Core Math Tools, available without charge to teachers. Core Math Tools is a suite of software tools organized into three families of software: algebra and functions, geometry and trigonometry, and statistics and probability. Core Math Tools was developed by the Core-Plus Mathematics Project (CPMP) under the guidance of Christian Hirsch and supported by funding from the National Science Foundation under grant DRL-1201917. http://www.nctm.org/resources/content.aspx?id=32702.

GeoGebra
GeoGebra is free multi-platform dynamic mathematics software for all levels of education that joins geometry, algebra, tables, graphing, statistics and calculus in one package. Graphics, algebra, and tables are dynamically connected. The GeoGebra home site contains not only the program but many additional resources, including sketches and a help network. GeoGebra can be downloaded at http://www.geogebra.org/cms/en/, and additional help can be found at http://www.geogebratube.org/.

Common Core State Standards for Mathematics (CCSSM)

Starting with the Standards
CCSSM is an important starting point for guiding curricular decisions. http://www.corestandards.org/Math.

Amplifying Meaning and Aligning the Standards
Bill McCallum, mathematician at the University of Arizona and lead member of the writing team that developed CCSSM, maintains a comprehensive website with many resources about the meaning of the standards and how to align teaching with them. http://ime.math.arizona.edu/commoncore.

Draft Progression Documents for the Standards
CCSSM was built on progressions that describe how a topic changes across a number of grade levels and are informed both by research on children's cognitive development and by the logical structure of mathematics. The progressions documents are all available online. http://ime.math.arizona.edu/progressions/.

Unpacking the Standards in Learning Trajectories
A website that unpacks CCSSM for kindergarten–grade 8 in eighteen learning trajectories, developed through the work of Jere Confrey, may help in understanding the elementary and middle school foundations for high school study of functions. TurnOnCCMath.net presents these trajectories in a set of graphics and narratives that connect the K–8 standards across the grade bands. The Generating Increased Science and Math Opportunities (GISMO) research team at North Carolina State University identified gaps in the standards and developed bridging standards to span those gaps. Clicking on the hexagons in the map at the website gives access to the bridging standards. http://turnonccmath.net.

Determining Alignment with the Standards
The website of the National Council of Supervisors of Mathematics offers resources to help with the implementation of CCSSM, kindergarten–grade 12. In particular, the site contains a tool that can help teachers determine the alignment of curriculum materials with CCSSM. Sponsored by the Council of Chief State School Officers and the National Governors Association and funded by Brookhill Foundation and Texas Instruments, the Curriculum Analysis Tool and the accompanying professional development material are available at http://www.mathedleadership.org/ccss/materials.html.

Mathematics Resources

Illuminations
The NCTM Illuminations project is part of the Verizon Thinkfinity program and presents a variety of standards-based resources, including lessons, activities, and hundreds of Web links. http://illuminations.nctm.org/.

Illustrative Mathematics

The Illustrative Mathematics website is designed to illustrate the range and types of mathematical work that students experience in a faithful implementation of the Common Core State Standards. The website also publishes other tools that support implementation of the standards—tasks and videos that show what *precision* means at various grade levels. http://www.illustrativemathematics.org/.

Math Forum

The Math Forum is a rich, free resource for teachers, kindergarten–grade 12. The site contains MathTools, a digital library of mathematics resources, including a catalog of materials and a discussion forum, to support mathematics instruction and learning. http://mathforum.org/mathtools/.

LearnZillion

LearnZillion is a learning platform that combines video lessons, assessments, and progress reporting for mathematics and English language arts. Each lesson highlights a Common Core standard (currently grades 3–9 in mathematics). Classroom teachers working with coaches created the lessons and materials on this site. The online lessons consist of PowerPoint slides with narratives lasting about 5 minutes. Each lesson addresses a specific standard and focuses on the conceptual understanding expected by that standard. The lessons are not intended to constitute a curriculum; however, they could serve as useful introductions or warm-up activities for a lesson or at the end of lessons. http://learnzillion.com/lessons.

Appendix 3
Tasks

This book examines rich tasks that can be used in the classroom to bring to the surface students' understandings and misunderstandings about functions. Many of these tasks are reproduced here, in the order in which they appear in the book, for the reader's personal or classroom use.

Function Wall

Goal
Understand the definition of a function.

Instructions to teacher: Place the letters A, B, C, and D along a wall (write each letter, large and bold, on a separate sheet of paper, taped to the wall, or on the board). Give students instructions on where to stand based on different characteristics. Explain that they will be extracting the definition of a function from this experience.

Question
Where do you stand?

1. Mode of traveling to school

    A. If you walked to school today, stand under A.

    B. If you rode your bike to school today, stand under B.

    C. If you drove or rode in a vehicle today, stand under C.

    D. If you got to school any other way, stand under D.

2. Time of travel to school

    A. If it took you between 0 and 10 minutes to get to school today, stand under A.

    B. If it took you between 10 and 30 minutes to get to school today, stand under B.

    C. If it took you between 30 minutes and an hour to get to school today, stand under C.

    D. If it took you more than an hour to get to school today, stand under D.

3. Grade in school

    A. If you're in seventh grade, stand under A.

    B. If you're in eighth grade, stand under B.

    C. If you're in ninth grade, stand under C.

    D. If you're in any other grade, stand under D.

Appendix 3

4. Clothing
    A. If you're wearing blue, stand under A.
    B. If you're wearing red, stand under B.
    C. If you're wearing black, stand under C.
    D. If you're wearing white, stand under D.

5. Birthday
    A. If you were born in January, stand under A.
    B. If you were born in February, stand under B.
    C. If you were born in March, stand under C.
    D. If you were born in April, stand under D.

6. Height
    A. If you're shorter than five feet tall, stand under A.
    B. If you're between than five feet and six feet tall, stand under B.
    C. If you're between six and seven feet tall, stand under C.
    D. If you're taller than seven feet tall, stand under D.

Function Finder

Goal
Understand the definition of a function.

Question
Which of these relationships are functions?

Decide whether each of the relationships below is a function. If it isn't a function, demonstrate how it fails to satisfy the definition of a function by showing elements of each set that do not "work." If it is a function, explain why, and also show several elements of each set.

Facebook user	password	student	hair color	students in our class	planet he/she lives on
state	letters in name	month	days in the month	days in month	month
date	temperature outside	password	Facebook user	any integer	double that integer

Appendix 3

Lake Depth

A park ranger measured the depth of the water in a lake at the same spot over a period of several weeks. He recorded the results in the table and graph shown in figures 1 and 2, respectively.

Change in days	Day	Depth	Change in depth
	7	15.29	
7	14	15.43	0.14
14	28	15.57	0.14
7	35	15.71	0.14
7	42	15.85	0.14

Fig. 1. The ranger's table of data on the water's depth

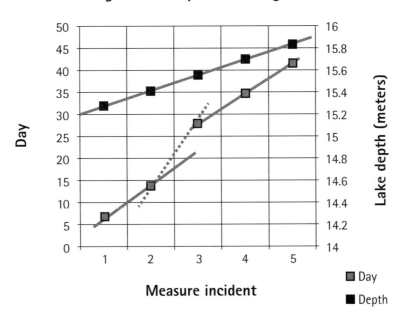

Fig. 2. The ranger's graph of the water depth data in his table

---

Adapted from Thomas J. Cooney, Sybilla Beckmann, and Gwendolyn M. Lloyd, *Developing Essential Understanding of Functions for Teaching Mathematics in Grades 9–12*, Essential Understanding Series (Reston, Va.: National Council of Teachers of Mathematics, 2010), p. 28.

Questions

1. In the table of data on the water depth (fig. 1), what does the gap in days mean about the measurement process?

2. In the graph (fig. 2), what is the rate of change for each of the lines?

3. What could explain the consistent depth measures in the lake depth data considering the gap in those measures? (That is, what happened to the lake?)

4. How does the gap in days in the table of data affect the covariance between the day and the lake depth?

5. Describe the meaning of the dashed line in the graph.

6. How does the overall average rate of change in lake depth compare to the average rate of change of the dashed line?

7. How would you describe the change in lake depth over time?

8. If the rate of change in lake depth remained constant (increase in depth per day), what depth would the park ranger have measured at time 3? At times 4 and 5?

Appendix 3

Possible responses from students

1. The rate of change in the depth of the lake is not constant. Students need to make a careful distinction between *change in depth* and *rate of change in depth*. Even though the values in the far right column (0.14) are all identical, the students cannot infer that the rate of change is constant. The rate of change in the depth of the water between the 7th and 14th days is 0.02 meters per day, whereas the rate of change in the depth of the water between the 14th and 28th days is 0.01 meters per day.

2. The rates of change for the three solid lines (starting at the top) are 0.14 meters per measurement, 7 days per measurement, and 7 days per measurement.

3. It took 7 days for the water's depth to increase 0.14 meters, from 15.29 meters to 15.43 meters. It then took 14 days for the water's depth to increase an additional 0.14 meters, from 15.43 meters to 15.57 meters. The smaller rate of increase in the depth of the water might reflect less rain during the time period or, perhaps, a release of water through a dam on the lake, lowering the rate of increase in the water level.

4. The covariance is not a constant rate. Although the change in the water's depth is constant over the five measurements, the corresponding change in days is not.

5. The slope of the dashed line is 14 days per measurement, which corresponds to the gap of 14 days between the second and third measurements.

6. The average rate of change is 0.016 meters per day. This means that if the total change in the depth of the water of 0.56 meters were evenly distributed over the 35 days in which the ranger measured it, he would have observed a change of 0.016 meters a day. However, the overall average change does not inform us about the day-to-day change. The average rate of change would still be 0.016 meters per day, even if the water depth remained constant over the first 41 days and then rose by 0.56 meters on the 42nd day.

7. The change in the lake's depth generally increases over the 35 days of measurement. The depth could fluctuate widely between the days on which it was measured; however, the rate of change is relatively constant for a week, decreases (but remains constant) for the next two weeks, and then increases to its original rate for the last week.

8. If the rate of change remained constant at the average rate of 0.016 meters per day, the depths that the ranger would have measured at time 3 (day 28), time 4 (day 35), and time 5 (day 42) would have been approximately $15.29 + 21(0.016) = 15.62$ meters, $15.29 + 28(0.016) = 15.74$ meters, and $15.29 + 35(0.016) = 15.85$ meters, respectively.

Parking Fees

Unlike many airports that have two different parking rates, one for short-term parking (for example, for passenger pickup) and the other for long-term parking, the local airport adjusts its parking fees automatically according to the time parked, to accommodate the different types of use. In the first 12 hours, parking costs $1 for each hour, and after 12 hours, parking costs $10 per day.

Problem

a. Write a piecewise function to describe parking costs at this airport.
b. Graph your piecewise function.

Questions

1. How do the short line segments in your graph differ from the longer segments?
2. What does the graph indicate that the fee is at 4 hours?
3. Note that the graph seems to overlap at 12 hours. What happens at the parking ticket booth then? What happens at 10 hours?
4. What happens from 12 to 24 hours? Why is this line segment a different length from the segments up to 12 hours and the segments after 24 hours? What does this line describe about the parking fee?
5. How does the fee change with the length of time that a vehicle is parked?
6. What would be the fee after x hours?
7. What would be the fee if you parked for 3 days? For 10 days?

Appendix 3

Possible responses from students

a. The piecewise function is $f(t) = \begin{cases} \lceil t \rceil & 0 \leq t \leq 12 \\ \left\lceil \dfrac{10t}{24} \right\rceil & t > 12 \end{cases}$, where t is time measured in hours.

b. The graph of the piecewise function is shown in figure 1.

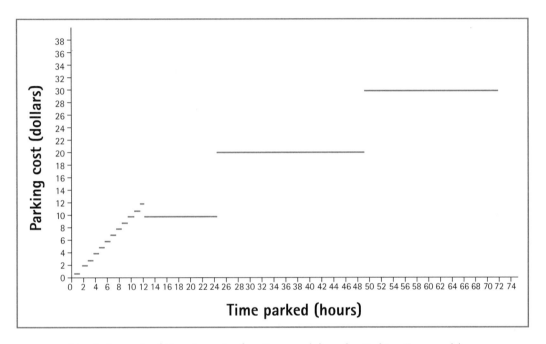

Fig. 1. A graph of the piecewise function modeling the Parking Fees problem

1. The short line segments correspond to an hourly parking rate, whereas the longer segments correspond to a daily parking rate.

2. The graph indicates that the fee at 4 hours is $4.

3. At 12 hours, the parking changes from an hourly rate to a daily rate. So if you park for t hours, $11 < t \leq 12$, you will be charged $12; if you park for 12 hours and 1 minute, you will be charged $10. If you park for 10 hours, the charge will be $10. If you park for 1 minute longer, the charge will be $11. At 10 hours (actually, immediately afterwards), the daily rate becomes cheaper than the hourly rate.

4. After 24 hours, the fee changes from $10 (one day) to $20 (two days). This line segment is shorter than the subsequent line segments (one-half as long) because customers would pay the daily rate of $10 only if they parked between 12 and 24 hours. After 24 hours,

the daily rates are constant over a 24-hour period.

5. The fee is a step-wise function of time. It increases $1 per hour for the first 12 hours, decreases to $10 for the next 12 hours, and then increases at a rate of $10 per day.

6. The fee could be determined by evaluating.

$$f(t) = \begin{cases} \lfloor t \rfloor & 0 \leq t \leq 12 \\ \left\lceil \dfrac{10t}{24} \right\rceil & t > 12 \end{cases}$$

for the given value of x.

7. The charge would be $30 for 3 days and $100 for 10 days.

Boat Rental

The community park has a small lake where visitors can rent paddle boats at $1 for 15 minutes, up to 2 hours. After 2 hours, the rate increases to $3 for 30 minutes.

Problem
Write the piecewise function to model this situation and graph the function.

Questions
1. What is the rental charge at 15 minutes? At 16 minutes? At 45 minutes?
2. If you had only $15, how long could you rent a boat?
3. How are the two lines in the graph the same, and how are they different?
4. How would the graph change if the rate change occurred at $t = 1$ hour?
5. How would the graph change if the fee were $2 for 15 minutes?
6. What is the rental fee at 3 hours?

Possible responses from students

Note that students can interpret this problem in more than one way. Assuming that the boat can be rented only in time intervals of 15 minutes (up to 2 hours) and 30 minutes (after 2 hours), they could interpret the function as the discrete function represented by the table of values in figure 1.

Time (minutes)	Cost (dollars)
15	1
30	2
45	3
60	4
75	5
90	6
105	7
120	8
150	11
180	14
210	17
240	20
270	23
300	26

Fig. 1. A table of values for the Boat Rental problem

If, however, after graphing these points, students then added trend lines, they could interpret the function representing the trend lines as

$$f(t) = \begin{cases} \dfrac{t}{15} & 0 \leq t \leq 120 \\ \dfrac{t}{10} - 4 & t > 120 \end{cases},$$

where *t* is time measured in minutes. The corresponding graph is given in figure 2.

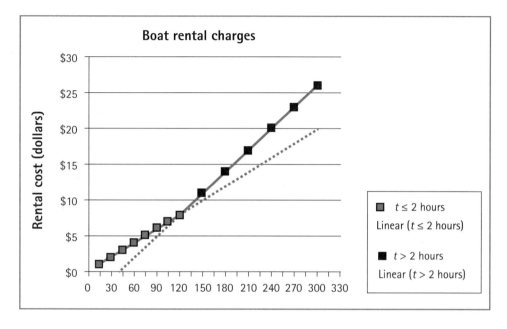

Fig. 2. A graph of the boat rental charges, with trend lines added

Further, students could interpret the function as a step function, where the amount of time that the boat is in use is rounded up to the nearest 15 minutes (if the time is less than or equal to 120 minutes) or 30 minutes (if the time is greater than or equal to 120 minutes). In that case, the function would be

$$f(t) = \begin{cases} \left\lceil \dfrac{t}{15} \right\rceil & 0 \leq t \leq 120 \\ \left\lceil \dfrac{3t}{30} \right\rceil & t > 120 \end{cases},$$

where *t* is time measured in minutes.
Responses to the questions follow:

1. The rental charge at 15 minutes is $1, at 16 minutes it is $2, and at 45 minutes it is $3.

2. With only $15, you could rent a boat for 180 minutes.

3. The two trend lines increase at a constant rate. However, the first line has a slope of $1/15$ ($1 per 15 minutes), while the second line has a slope of $3/30 = 1/10$.

4. The second trend line would be shifted to the left by 60 units.

5. The slope of the first line would double from $1/15$ to $2/15$.

6. The rental fee at 3 hours is $8 for the first two hours and $6 for the third hour, or a total of $14. The fee could also be found by looking at the point corresponding to $t =$ 180 minutes on the graph, or $14.

Appendix 3

Falling Object

An object starts falling at a height of 400 feet. The graph in figure 1 models the function.

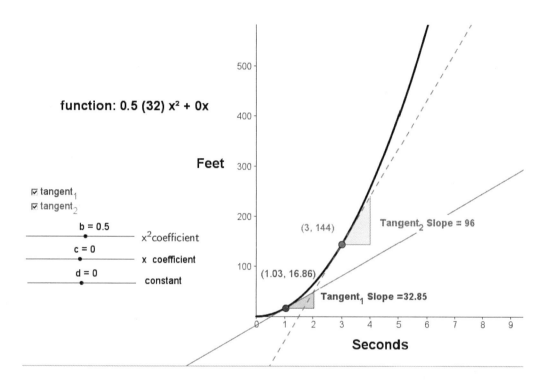

Fig. 1. The graph of the falling object

Questions

1. What are reasonable values for the domain and range of this function, and why?
2. What does the value 32 represent in the equation of the function?
3. What is the direction of the object? How is this shown in the graph?
4. What is the distance that the object has fallen in 1 second? In 2 seconds?
5. How long does the object take to hit the ground? How is this represented in the graph?
6. What does the shape of the graph indicate about the rate of fall?

7. What does the slope mean at time = 2 seconds? At time = 3 seconds? What will it be at time = 5 seconds?

8. In the graph of distance fallen as a function of time (fig. 1), can you find a pattern in the instantaneous slope and the time?

9. What does a slope of zero mean in this situation?

10. What is the maximum rate of change?

11. What does the slope mean in terms of the story?

12. If the object started at 500 feet, how would the graph change?

Appendix 3

Possible responses from students

1. The domain consists of the real numbers between 0 and 5, since it takes 5 seconds for the object to hit the ground. The object starts at a height of 400 feet, so the range consists of the real numbers between 0 and 400.

2. The value 32 represents the acceleration of the object due to gravity: 32 feet per second per second.

3. The object falls toward the earth. The graph is not a picture or other sort of image of the falling object, but rather it shows the distance that the object has fallen. Because the distance that the object falls increases over time, we know that the object is falling to earth.

4. The object falls 16 feet in the first second and has fallen 64 feet after 2 seconds.

5. The time that the object takes to hit the ground is the value of the domain that is paired with 400 in the range—namely, 5 seconds.

6. The shape of the graph indicates that the rate at which the object falls increases over time.

7. The slope of the graph at time = 2 seconds and at time = 3 seconds (which is the slope of the corresponding tangent lines) corresponds to the rate at which the object is falling at those times. The object will hit the ground at 5 seconds.

8. At t seconds, the slope of the graph will be $32t$. This corresponds to the object's fall at a rate of $32t$ feet per second.

9. A slope of zero corresponds to the object falling at a constant rate.

10. Because the rate at which the object falls is always increasing, the maximum rate of change would be at 5 seconds when the object hits the ground, or 160 feet per second.

11. The slope corresponds to the rate at which the object is falling at a given time.

12. The graph would not change. However, the upper limits of the domain and range would change to approximately 5.6 seconds and 500 feet, respectively.

Sliding Ladder

An 18-foot ladder is leaning against a house when its base starts to slide away. By the time the base is 12 feet from the house, the base is moving at a rate of 5 ft/sec. How fast is the top of the ladder sliding down the wall then?

Figure 1 shows sketches that represent the initial position and two later positions of the ladder in its slide.

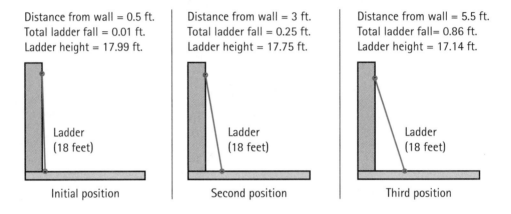

Fig. 1. The ladder at its initial position and two subsequent positions as it slides down the wall

Questions

1. Will the top or the bottom of the ladder move faster?
2. Will the top and the bottom of the ladder move at the same rate? When?
3. If the bottom of the ladder moves one foot, how far will the top slide down the wall?
4. Will the bottom of the ladder always move faster than the top?
5. How can you compare the change in the movement of the top of the ladder with the change in the movement at the bottom of the ladder?

    Adapted from Stephen Monk, "Students' Understanding of a Function Given by a Physical Model," in *The Concept of Function: Aspects of Epistemology and Pedagogy,* edited by Guershon Harel and Ed Dubinsky (MAA Notes, no. 25 [1992]), pp. 175–94.

Appendix 3

Possible responses from students

Students may find it useful to simulate the situation with interactive software and make a table of values, as in figure 2. This table was generated with tools available in GeoGebra (http://www.geogebra.org/cms/en). Using GeoGebra, students can move the base of the ladder 1 foot at a time, as in figure 1, while also recording the movement of the top of the ladder. In the table—

- Base = horizontal distance (ft.) of the bottom of the ladder from the wall;
- Fall = vertical distance (ft.) that the ladder has fallen; and
- Height = vertical distance (ft.) of the top of the ladder from the ground as it falls.

Referring to the table helps students in responding to the questions.

Base	Fall	Height
1	0.03	17.97
2	0.11	17.89
3	0.25	17.75
4	0.45	17.55
5	0.71	17.29
6	1.03	16.97
7	1.42	16.58
8	1.88	16.12
9	2.41	15.59
10	3.03	14.97
11	3.75	14.25
12	4.58	13.42
13	5.55	12.45
14	6.69	11.31
15	8.05	9.95
16	9.75	8.25
17	12.08	5.92
17.99	17.40	0.60

Fig. 2. A table of values for the Sliding Ladder problem

1. At the beginning, the bottom of the ladder moves faster.
2. From step 12 to step 13, the changes in the bottom and top of the ladder are the same (both are 0.97 ft.).
3. How far the top of the ladder falls varies. From step 1 to step 2 it falls 0.08 feet, from step 10 to step 11 it falls 0.72 feet, and from step 15 to step 16 it falls 1.7 feet.
4. No, the bottom of the ladder will not always move faster than the top. For example, from step 1 to step 2, the top falls only 0.08 feet, but from step 15 to step 16 it falls 1.7 feet.
5. The rate at which the top of the ladder falls increases over time, and eventually the top falls at a faster rate than the bottom of the ladder.

Appendix 3

Water's Height in a Trough

A water trough (see fig. 1) is 10 inches wide at the base and 10 inches deep. It has a right angle at the back and a 135-degree angle at the front. Water is filling the trough at a constant rate. When the water reaches a height of 4 inches, the width of its surface on the end of the trough is 14 inches.

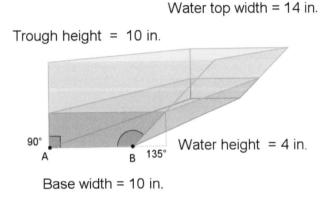

Fig. 1. A water trough with a right trapezoidal cross section

Problem

a. What is the function that shows the water's width, given the trough pictured, with a height of 10 inches and a base width of 10 inches? What would be the function for a base width of 5 inches? Of 15 inches? Of 20 inches? Of w_0 inches?

b. Graph the functions that you found for base width values of 5, 10, 15, and 20.

c. What is the function that shows the rate of change of the water's width, given its height?

Questions

1. In the functions for base widths of 5, 10, 15, and 20 inches, what do the four functions describe together? How does changing the width of the trough change the function?

2. Why are the lines for the four functions parallel? Describe the parallelism in terms of the rising water.

3. What happens at $h = 0$?

4. What is changing in the functions? What stays the same? Describe what changes and what stays the same in terms of the water in the trough and in terms of the graph.

5. How would the function change if the base were zero?

6. How would the function change if both sides of the trough were set at 135 degrees? At 90 degrees?

7. What are the limits of the domain and range for these functions?

Appendix 3

Possible responses from students

a. $f_2(h) = h + 10$; $f_1(h) = h + 5$; $f_3(h) = h + 15$; $f_4(h) = h + 20$ (named as in graph in [b] below); $f(h) = h + w_0$

b. Figure 2 shows the graphs for base width values of 5, 10, 15, and 20:

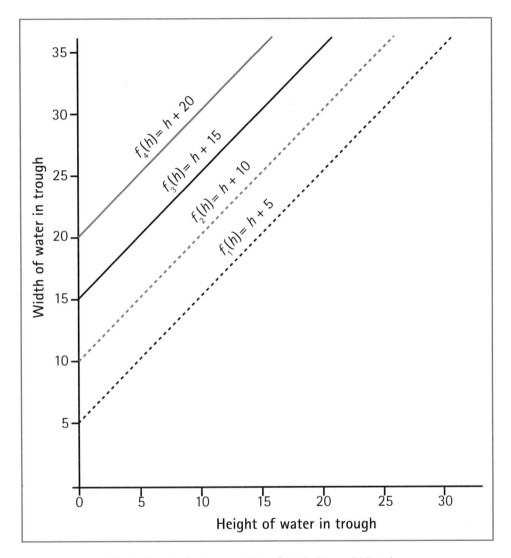

Fig. 2. Graphs for base widths of 5, 10, 20, and 20 inches

c. The function that shows the rate of change of the water's width, given its height, is discussed below, after the possible response to question 7.

1. The graphs of these four functions show that changing the base width of the trough affects the height-width function in a predictable way. The base width is the constant w_0 in the function $f(h) = h + w_0$.

2. The slopes of the lines correspond to the rate at which the width of the water changes in relation to the height of the water. Because this rate is constant for any given base width, the slopes of the lines are all equal, and the lines are parallel.

3. At $h = 0$, no water has been poured into the trough, so the width is equal to the base width.

4. In a given function, the width $f(h)$ is changing as the height h changes. The ratio of the width to height of the water (as represented by the slope of the line) does not change. Additionally, in a change from one function to another, the base width changes (as represented by the y-intercept of the graph).

5. If the base width were zero, the cross-section of the trough would be an isosceles right triangle, and the width and the height would be equal ($f(h) = h$), which corresponds to the general equation $f(h) = h + w_0$ with $w_0 = 0$.

6. If both sides of the trough were set at 135 degrees, the general function would be $f(h) = 2h + w_0$. If both sides of the trough were set at 90°, the width would remain constant, and the general function would be $f(h) = w_0$.

7. For the original problem, the domain is the set of real numbers between 0 and 10, and the range is the set of real numbers between 10 and 20.

The question posed in (c) was, What is the function that shows the rate of change of the water's width, given its height? Because an increase in height of one foot results in an increase in width of 1 foot, the desired function is $g(x) = 1$.

Appendix 3

Water's Surface Area on the End of a Trough

A water trough (see fig. 1) has a base of 10 inches and a height of 10 inches. Both the front and the back side of the trough form a 135-degree angle at the base of the trough. Water is filling the trough at a constant rate and has reached a height of 4 inches and a width of 18 inches across the end of the trough.

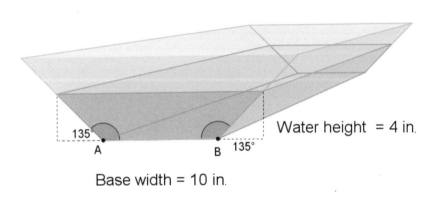

Fig. 1. A water trough with an isosceles trapezoidal cross section

Problem

a. What is the function that shows the surface area of the water on the end of the trough, given a height of 10 inches and a base width of 10 inches, as pictured in figure 1?

b. What is the function that shows the surface area of the water on the end of the trough, given a height of 10 inches, if the base width is 6 inches? If the base width is 0 inches?

c. Graph the three functions that you found in (a) and (b).

d. How does the surface area of the water on the end of the trough change with the water's height?

Questions

1. What do the three functions that you found in (a) and (b) describe together?

2. How does changing the width of the base of the trough change the function?

3. What happens at $h = 0$?

4. What is the value for each of the functions at $h = 1$? What does this mean in terms of water in the trough?

5. What is changing in the functions? What stays the same? Describe what changes and what stays the same in terms of the water in the trough and in terms of the graph.

6. How would the function change if both sides of the trough were changed?

7. What are the limits of the domain and range for these functions?

Appendix 3

Possible responses from students

a. If the width of the water is $2h + 10$ and the height of the water is h, then the formula for finding the area of a trapezoid gives a surface area function of $a_1(h) = h^2 + 10h$.

b. In general, the function is given by $a(h) = h^2 + hb$, where b = base width. So, the function becomes $a_2(h) = h^2 + 6h$ for a base width of 6 inches, and $a_3(h) = h^2$ for a base width of 0 inches.

c. Figure 2 shows the graphs of the functions for base width values of 0, 6, and 10.

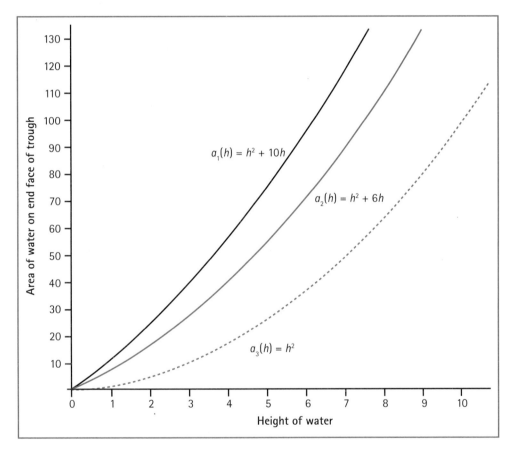

Fig. 2. Graphs of the functions for base widths of 0, 6, and 10 inches

d. How the surface area of the water on the end of the trough changes with the water's height is discussed below, after the possible response to question 7.

1. The graphs of these three functions show that changing the base width of the trough affects the height-surface area function in a predictable way, although not as uniformly as the height-width function.

2. The base width is the constant b in the function $a(h) = h^2 + hb$.

3. At $h = 0$, no water has been poured into the trough, so the surface area is 0 square inches.

4. At $h = 1$, the water is 1 inch deep. The corresponding surface areas are $a_1(1) = 11$ square inches, $a_2(1) = 7$ square inches, and $a_3(1) = 1$ square inch.

5. In a given function, the water's height h, width $2h + b$, and surface area $a(h)$ are all changing. The structure of the equation $a(h) = h^2 + bh$ remains the same, as does the constant b, which represents the base width. The change in the height of the water affects a change in the surface area in two different ways—both in the quadratic and in the linear terms of the formula. This results in the surface area increasing at an increasing rate.

6. How the function would change if both sides of the trough were changed depends on the trough's new shape. As the cross-sectional shape of the trough changes, the formula for computing the surface area also changes. The rate at which the surface area changes is dependent on the shape of the trough.

7. For these functions, the domain is the set of real numbers between 0 and 10, and the range is the set of real numbers between 0 and 200.

The problem posed in (d) was, How does the surface area of the water on an end of the trough change with the water's height? As the height of the water increases, the surface area increases at an increasing rate. In fact, this rate turns out to be $2h + b$, where b is the base width. This rate seems reasonable, since as the height of the water increases by a very small amount, Δh, a correspondingly small very thin pseudo-rectangle of area $\Delta h(2h + b)$ is added to the surface area.

Appendix 3

Velocity of One Car

The graph in figure 1 shows the velocity of a car over a period of 5 minutes.

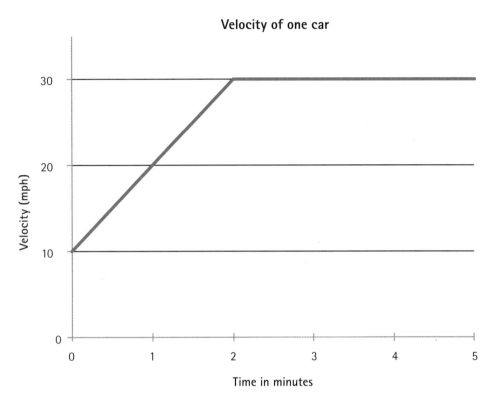

Fig. 1. Velocity graph of one car. Adapted from Monk (2003), p. 251.

Questions

1. What do the two line segments in the graph indicate? Compare the velocity represented by the oblique (slanted) segment with the velocity represented by the horizontal segment.

2. People may misinterpret the meaning of the straight line segments in the graph. Do these straight segments indicate no change? How does the oblique segment differ from

---

Adapted from Stephen Monk, "Representation in School Mathematics: Learning to Graph and Graphing to Learn," in *A Research Companion to Principles and Standards for School Mathematics*, edited by Jeremy Kilpatrick, W. Gary Martin, and Deborah Schifter (Reston, Va.: National Council of Teachers of Mathematics, 2003), pp. 250–62.

the horizontal segment in meaning? What would the car have to do for these lines to be more "wavy"?

3. How would you compare the car's action at $t = 0$ and its action at $t = 3$? What does it mean that the graph starts at a velocity of 10 miles per hour? Can a car be going 10 miles per hour at time $t = 0$? What can you say about the velocity of the car at $t = 6$?

4. What happens at $t = 2$? Does the car stop? Write a short story (just a few lines) describing what it would be like if you were in the car represented by this graph.

Possible responses from students

1. The two line segments represent the velocity at which the car is moving over time. The slanted segment between 0 and 2 minutes indicates that the car is accelerating at a constant rate of 10 miles per hour per minute, from 10 miles per hour to 30 miles per hour. The horizontal line segment indicates that the car travels at a constant velocity of 30 miles per hour for the final 3 minutes.

2. The car would have to alternately speed up and then slow down to make the lines more wavy.

3. At the moment that the velocity of the car is initially recorded, it is traveling at 10 miles per hour. (It would have to have had a head start.) It is unknown what the speed of the car is at $t = 6$.

4. If you were in the car, you would feel the car smoothly speeding up for 2 minutes and then staying at a constant velocity for the next 3 minutes.

Putting Essential Understanding of Functions into Practice in Grades 9–12

Comparing Two Cars, Given Distance

The graph in figure 1 represents the distance, from a given location, of two cars, A and B, traveling in the same direction on the same road.

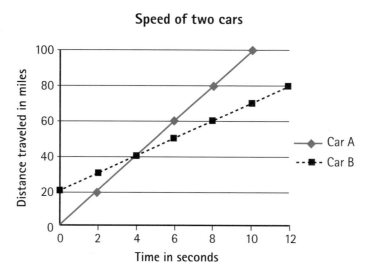

Fig. 1. Distance traveled by two cars

Questions

1. Describe what is happening at $t = 2$. Which car is going faster? How can you tell? Which car has gone farther? How far apart are the two cars?

2. Are the cars ever moving at the same speed? At what time? Explain how the graph represents this event.

3. Describe location of the cars at $t = 0$. What can be said about the speeds of the cars at this time?

4. Does one car ever pass the other? At what time?

5. Have the two cars traveled the same distance at any point in time? If so, when?

Adapted from Lillian McDermott, Mark L. Rosenquist, and Emily H. van Zee, "Student Difficulties in Connecting Graphs and Physics: Example from Kinematics," *American Journal of Physics* 55, no. 6 (1987): 503–13.

Appendix 3

Possible responses from students

1. At $t = 2$, car A is traveling at a greater speed than car B, since the slope of the line representing car A's speed is greater than the slope of the line representing car B's speed. Car B has traveled 10 miles, and is currently 30 miles from the starting point. Car A has traveled 20 miles from the starting point. So car B is 10 miles ahead of car A.

2. Because the slopes of the graphs for the two cars are never equal, their speeds can never be the same.

3. Car B does not start at the same place on the road as car A; car B has a 20-mile head start. Car A's speed is a constant.

4. At $t = 4$ both cars are 40 miles away from the starting point, and car A will pass car B at that time.

5. The only time in the graph that the cars travel the same distance is at $t = 0$.

Comparing Two Cars, Given Speed

Two cars, A and B, start at the same time and place and travel for one hour. The trips for car A and car B are shown in the graph in figure 1.

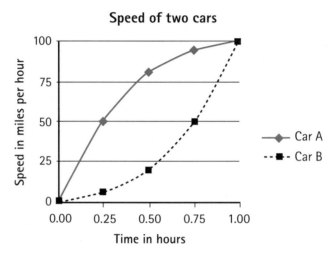

Fig. 1. A graph of the speed of two cars

Questions

1. Which car is traveling faster at $t = 0.75$ hour?
2. Describe what happens at $t = 1$.
3. Describe the relationship between the cars at $t = 0.25$ ($1/4$ hour).
4. Which car travels farther?
5. What is the closest that the cars come to each other?

Adapted from John Clement, "The Concept of Variation and Misconceptions in Cartesian Graphing," *Focus On Learning Problems in Mathematics* 11 (Winter/Spring 1989), pp. 77–87, and Stephen Monk, "Students' Understanding of Functions in Calculus Courses," *Humanistic Mathematics Network Journal* 9 (February 1994), pp. 21–27.

Appendix 3

Possible responses from students

1. Because the line representing car A's speed is above the line representing car B's speed, car A is traveling faster than car B at $t = 0.75$ hours.

2. At $t = 1$, the cars are traveling at the same speed.

3. The graph indicates that at $t = 0.25$, car A is traveling 50 miles per hour and car B is moving at about 6 miles per hour. So car A is traveling at a rate that is more than 8 times that of car B and is far ahead of car B (and pulling farther ahead at all times, except at the end). Students should be able to see that Car A's speed is increasing at a steady rate until about 0.25 hours, after which it is still increasing but at a decreasing rate. By contrast, Car B's rate is increasing steadily until about 0.25 hours, at which point it begins to increase at an increasing rate.

4. Because the line representing car A's speed is always above the line representing car B's speed (except at the end), car A has traveled a much greater distance than car B.

5. The only time that the two cars are together is at the beginning, $t = 0$.

Two Walkers

Amanda and Joe are standing next to each other and start walking at the same time, along parallel lines. Amanda starts by taking big steps, and each step grows smaller as she walks. Joe starts with small steps, and each step gets larger as he walks.

The table in figure 1 shows Amanda's and Joe's steps and step sizes. Describe Amanda's and Joe's trips.

Step number (s)	Step size (ft.) Amanda	Joe
1	3.00	0.0
2	2.75	0.5
3	2.50	1.0
4	2.25	1.5
5	2.00	2.0
6	1.75	2.5
7	1.50	3.0
8	1.25	3.5
9	1.00	4.0
10	0.75	4.5

Fig. 1. A table showing step numbers and sizes for Amanda and Joe. From Monk (2003, p. 253).

Questions

1. What do Amanda's and Joe's trips look like?
2. What happens at the fifth step, $s = 5$?
3. What happens to Joe at $s = 1$?
4. Who travels farther?

Adapted from Stephen Monk, "Representation in School Mathematics: Learning to Graph and Graphing to Learn," in *A Research Companion to Principles and Standards for School Mathematics*, edited by Jeremy Kilpatrick, W. Gary Martin, and Deborah Schifter (Reston, Va.: National Council of Teachers of Mathematics, 2003), pp. 250–62.

5. Who is ahead at $s = 7$?

6. At what time have Amanda and Joe traveled the same distance? Does Joe ever catch Amanda?

Putting Essential Understanding of Functions into Practice in Grades 9–12

Possible responses from students

1. Figures 2 and 3 show graphs of two types that students may create:

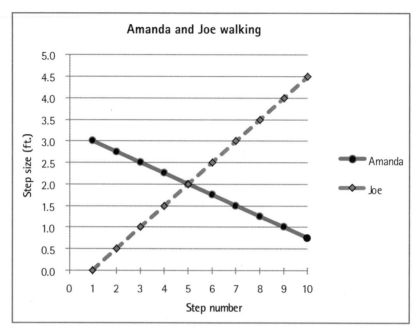

Fig. 2. Steps by step size

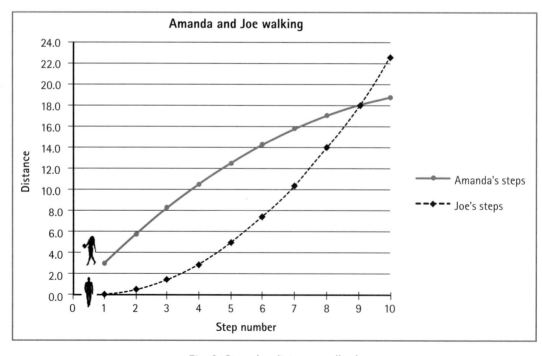

Fig. 3. Steps by distance walked

Appendix 3

2. At $s = 5$, Amanda's and Joe's steps are the same size (2 feet). However, at this point, Amanda has walked 12.5 feet, and Joe has walked 5 feet.

3. Both Amanda and Joe are at the same point at $s = 0$. At $s = 1$, Joe takes a step of 0 feet—that is, he does not move. Amanda takes a step of 3 feet, and this results in a 3-foot head start for her.

4. Joe walks farther. He walks a total of 22.5 feet, and Amanda walks a total of 18.75 feet.

5. Amanda is ahead at $s = 7$. She has walked 15.75 feet, and Joe has walked 10.5 feet. However, Joe is catching up. At $s = 8$, Amanda has walked 17 feet, and Joe has walked 14 feet. The difference in distance has narrowed from 5.25 feet to 3 feet.

6. Amanda and Joe have both traveled the same distance (18 feet) at $s = 9$.

References

Akgün, Levent, and M. Emin Özdemir. "Students' Understanding of the Variable as General Number and Unknown: A Case Study." *The Teaching of Mathematics* 9 (March 2006): 45–51.

Bayazit, Ibrahim. "Selection and Resolution of Function Problems and Their Effects on Student Learning." *Educational Research and Reviews* 6 (November 2011): 906–18.

Borba, Marcelo C., and Jere Confrey. "A Student's Construction of Transformations of Functions in a Multiple Representational Environment." *Educational Studies in Mathematics* 31 (October 1996): 319–37.

Carlson, Marilyn, Sally Jacobs, Edward Coe, Sean Larsen, and Eric Hsu. "Applying Covariational Reasoning While Modeling Dynamic Events: A Framework and a Study." *Journal for Research in Mathematics Education* 33 (November 2002): 352–78.

Clement, John. "The Concept of Variation and Misconceptions in Cartesian Graphing." *Focus on Learning Problems in Mathematics* 11 (Winter/Spring 1989): 77–87.

Clement, Lisa L. "What Do Students Really Know about Functions?" *Mathematics Teacher* 94 (December 2001): 745–48.

Common Core State Standards Writing Team. Draft Progressions on High School Algebra and Functions. Tools for the Common Core, 2012. http://commoncoretools.me/2012/12/04/draft-progressions-on-high-school-algebra-and-functions/.

Cooney, Thomas J., Sybilla Beckmann, and Gwendolyn M. Lloyd. *Developing Essential Understanding of Functions for Teaching Mathematics in Grades 9–12*. Essential Understanding Series. Reston, Va.: National Council of Teachers of Mathematics, 2010.

De Bock, Dirk, Wim Van Dooren, Dirk Janssens, and Lieven Verschaffel. "Improper Use of Linear Reasoning: An In-Depth Study of the Nature and the Irresistibility of Secondary School Students' Errors." *Educational Studies in Mathematics* 50 (August 2002): 311–34.

Dotson, Geraldine Ting. "Collegiate Mathematics Students' Misconceptions of Domain and Zeros of Rational Functions." PhD diss., University of Kansas, 2009.

Dougherty, Barbara J. "Access to Algebra: A Process Approach." In *The Future of the Teaching and Learning of Algebra*, edited by Helen Chick, Kaye Stacey, Jill Vincent, and John Vincent, pp. 207–13. Victoria, Australia: University of Melbourne, 2001.

Even, Ruhama. "Subject-Matter Knowledge and Pedagogical Content Knowledge: Prospective Secondary Teachers and the Function Concept." *Journal for Research in Mathematics Education* 24 (March 1993): 94–116.

Grossman, Pamela. *The Making of a Teacher.* New York: Teachers College Press, 1990.

Gür, Hulya, and Baflak Barak. "The Erroneous Derivative Examples of Eleventh Grade Students." *Educational Sciences: Theory and Practice* 7 (January 2007): 473–80.

Hayakawa, S. I. *Language in Thought and Action.* London: G. Allen and Unwin, 1952.

Hill, Heather C., Brian Rowan, and Deborah Loewenberg Ball. "Effects of Teachers' Mathematical Knowledge for Teaching on Student Achievement." *American Educational Research Journal* 42 (Summer 2005): 371–406.

Kaput, James J. "Linking Representations in the Symbolic Systems of Algebra." In *Research Agenda for Mathematics Education: Research Issues in the Learning and Teaching of Algebra,* edited by Sigrid Wagner and Carolyn Kieran, 167–94. Reston, Va.: National Council of Teachers of Mathematics, 1989.

Kuchemann, Dietmar. "Children's Understanding of Numerical Variables." *Mathematics in School* 7 (September 1978): 23–26.

Leinhardt, Gaea, Orit Zaslavsky, and Mary Kay Stein. "Functions, Graphs, and Graphing: Tasks, Learning, and Teaching." *Review of Educational Research* 60 (Spring 1990): 1–64.

Magnusson, Shirley, Joseph Krajcik, and Hilda Borko. "Nature, Sources, and Development of Pedagogical Content Knowledge for Science Teaching." In *Examining Pedagogical Content Knowledge,* edited by Julie Gess-Newsome and Norman G. Lederman, pp. 95–132. Dordrecht, The Netherlands: Kluwer Academic, 1999.

McDermott, Lillian, Mark L. Rosenquist, and Emily H. van Zee. "Student Difficulties in Connecting Graphs and Physics: Example from Kinematics." *American Journal of Physics* 55, no. 6 (1987): 503–13.

McKenzie, Danny L., and Michael J. Padilla. "The Construction and Validation of the Test of Graphing in Science (TOGS)." *Journal of Research in Science Teaching* 23 (October 1986): 571–79.

Monk, Stephen. "Students' Understanding of a Function Given by a Physical Model." In *The Concept of Function: Aspects of Epistemology and Pedagogy,* edited by Guershon Harel and Ed Dubinsky, pp. 175–94. MAA Notes Series, no. 25. Washington, D.C.: Mathematical Association of America, 1992.

———. "Students' Understanding of Functions in Calculus Courses." *Humanistic Mathematics Network Journal* 9 (February 1994): 21–27.

———. "Representation in School Mathematics: Learning to Graph and Graphing to Learn." In *A Research Companion to Principles and Standards for School Mathematics*, edited by Jeremy Kilpatrick, W. Gary Martin, and Deborah Schifter, pp. 250-62. Reston, Va.: NCTM, 2003.

National Council of Teachers of Mathematics (NCTM). *Principles and Standards for School Mathematics*. Reston, Va.: NCTM, 2000.

National Governors Association Center for Best Practices and Council of Chief State School Officers (NGA Center and CCSSO). *Common Core State Standards for Mathematics. Common Core State Standards (College- and Career-Readiness Standards and K–12 Standards in English Language Arts and Math)*. Washington, D.C.: NGA Center and CCSSO, 2010. http://www.corestandards.org.

Philipp, Randolph A. "The Many Uses of Algebraic Variables." *Mathematics Teacher* 85 (October 1992): 557-61.

Popham, W. James. "Defining and Enhancing Formative Assessment." Paper presented at the CCSSO State Collaborative on Assessment and Student Standards FAST meeting, Austin, Tex., October 10-13, 2006.

Russell, Michael, Laura O'Dwyer, and Helena Miranda. "Diagnosing Students' Misconceptions in Algebra: Results from an Experimental Pilot Study." *Behavior Research Methods* 41 (May 2009): 414-24.

Schoenfeld, Alan H., and Abraham Arcavi. "On the Meaning of Variable." *Mathematics Teacher* 81 (September 1988): 420-27.

Shulman, Lee S. "Those Who Understand: Knowledge Growth in Teaching." *Educational Researcher* 15, no. 2 (1986): 4-14.

———. "Knowledge and Teaching." *Harvard Educational Review* 57, no. 1 (1987): 1-22.

Vinner, Shlomo. "Concept Definition, Concept Image and the Notion of Function." *International Journal of Mathematical Education in Science and Technology* 14, no. 3 (1983): 293-305. http://dx.doi.org/10.1080/0020739830140305.

Vinner, Shlomo, and Tommy Dreyfus. "Images and Definitions for the Concept of Function." *Journal for Research in Mathematics Education* 20 (July 1989): 356-66.

Wiliam, Dylan. "Keeping Learning on Track: Classroom Assessment and the Regulation of Learning." In *Second Handbook of Research on Mathematics Teaching and Learning*, edited by Frank K. Lester, Jr., pp. 1053-98. Charlotte, N.C.: Information Age; Reston, Va.: National Council of Teachers of Mathematics, 2007.

Yinger, Robert J. "The Conversation of Teaching: Patterns of Explanation in Mathematics Lessons." Paper presented at the meeting of the International Study Association on Teacher Thinking, Nottingham, England, May 1998.